MW01267693

ADDITIONAL SKILL AND DRILL MANUAL

JAMES J. BALL
Indiana State University

INTERMEDIATE ALGEBRA

EIGHTH EDITION

Margaret L. Lial
American River College

Books
Books
9412080660 1
$2.49

John Hornsby
University of New Orleans

Terry McGinnis

PEARSON
Addison Wesley

Boston San Francisco New York
London Toronto Sydney Tokyo Singapore Madrid
Mexico City Munich Paris Cape Town Hong Kong Montreal

Reproduced by Pearson Addison-Wesley from electronic files supplied by the author.

Copyright © 2006 Pearson Education, Inc.
Publishing as Pearson Addison-Wesley, 75 Arlington Street, Boston, MA 02116.

All rights reserved. No part of this publication may be reproduced, stored in a retrieval system, or transmitted, in any form or by any means, electronic, mechanical, photocopying, recording, or otherwise, without the prior written permission of the publisher. Printed in the United States of America.

ISBN   0-321-33166-4

    7 8 9 BRR 08 07

CONTENTS

Chapter 1

REVIEW OF THE REAL NUMBER SYSTEM

1.1   Basic Concepts

Objective 1   Write **sets using set notation.**

Write whether each statement is **true** *or* **false.**

1. All natural numbers are counting numbers.

2. A number that varies in an expression is called a constant.

3. There is only one way to describe any set in set-builder notation.

4. The members of a set may also be called the elements of the set.

5. $\{x|x$ is a month beginning with the letter J$\}$ defines the set $\{$June, July$\}$.

6. The set $\{4, 8, 12, 16, \ldots\}$ is defined by $\{y \mid y$ is a whole number that is divisible by 4$\}$.

Write each set by listing its elements.

7. $\{a, b, c, d, e\}$

8. $\varnothing$

9. $\{w|w$ is a natural number greater than 6 but less than 11$\}$

10. $\{z|z$ is a state beginning with the letter O$\}$

11. $\{x|x$ is one of the last three presidents of the United States$\}$

12. $\{z|z$ is an odd natural number less than 9$\}$

13. $\{y|y$ is a natural number divisible by 5$\}$

Objective 2   **Use number lines.**

Graph the elements of each set on a number line.

14. $\{1, 3, 5, 7\}$

15. $\{-3, -2, -1, 0, 1\}$

16. $\{0, 2, 4, 6, 8\}$

17. $\{-3, 7, 10\}$

18. $\left\{\dfrac{1}{2}, \dfrac{3}{2}, \dfrac{5}{2}, \dfrac{7}{2}\right\}$

19. $\left\{-\dfrac{1}{3}, 0, \dfrac{1}{3}, \dfrac{2}{3}\right\}$

20. $\left\{-4, -2.5, 0, \sqrt{9}\right\}$

21. $\left\{-\dfrac{9}{2}, -\sqrt{4}, 1.5, 3\right\}$

22. $\left\{-4, -\dfrac{3}{2}, \dfrac{1}{2}, 5\right\}$

23. $\left\{3, \dfrac{3}{2}, 1, \dfrac{3}{4}\right\}$

Objective 3 **Know the common sets of numbers.**

Which elements of the set $\left\{-4, -\sqrt{2}, -\frac{1}{3}, 0, \frac{4}{5}, \sqrt{7}, 5\right\}$ *are elements of the following sets?*

24. Natural numbers   **26.** Irrational numbers   **28.** Rational numbers

25. Integers   **27.** Whole numbers   **29.** Real numbers

Write whether each statement is **true** *or* **false.**

30. All rational numbers are real numbers.

31. All integers are whole numbers.

32. Some irrational numbers are rational numbers.

33. All whole numbers are counting numbers.

34. All integers are rational numbers.

35. Some rational numbers are whole numbers.

36. Some irrational numbers are whole numbers.

37. All irrational numbers are real numbers.

Objective 4 **Find additive inverses.**

Find each additive inverse.

38. 0

39. −2

40. 3

41. 2

42. −π

43. $\sqrt{2}$

44. −.75

45. 2.5

46. $-\dfrac{5}{2}$

47. $-\dfrac{6}{5}$

48. $\dfrac{10}{17}$

49. $\dfrac{8}{9}$

Objective 5 **Use absolute value.**

Find the value of each expression.

50. $|12|$

51. $|-7|$

52. $-|8|$

53. $-|-8|$

54. $|0|$

55. $|6|-|-1|$

56. $\left|7\right|+\left|-2\right|$

57. $\left|-5\right|-\left|-2\right|$

58. $\left|-7\right|+\left|-8\right|$

59. $\left|-12\right|-\left|2\right|-\left|-6\right|$

60. $\left|-2\right|+\left|17\right|-\left|-4\right|$

61. $\left|-10\right|-\left|-1\right|-\left|-4\right|$

Objective 6   **Use inequality symbols.**

*Use a number line to identify each inequality as **true** or **false**.*

62. $3 > 5$

63. $-2 < 7$

64. $8 > -4$

65. $-3 < -5$

66. $0 > -1$

67. $-8 \le -9$

68. $-1 > -5$

69. $5 \ge 5$

70. $-8 \ge -8$

Use inequality symbols to write each statement.

71. 21 is greater than 13.

72. -8 is less than 2.

73. -6 is less than or equal to -3.

74. 0 is greater than or equal to x.

*Simplify each side of the statement to tell whether the resulting statement is **true** or **false**.*

75. $-11 \ge 8 + 2$

76. $\dfrac{17}{5} > -\left|-3\right|$

77. $-\left|7\right| \le \left|-8\right|$

78. $-\left|7-4\right| \le -4$

79. $8 \cdot 0 \ge 3 \cdot 9$

80. $3 + 5 \le \left|-7\right|$

81. $-\left|-2\right| \ge \left|-3\right|$

82. $4(7) - 3 \ge 3(5)$

83. $4 \cdot 0 \ge 3 \cdot 0$

84. $-4 \ge -4$

1.1 Mixed Exercises

*Write whether each statement is **true** or **false**.*

85. The empty set is written $\{0\}$.

86. Examples of signed numbers are -8, $-\sqrt{3}$, and 5.

87. The coordinate of a point is the graph of the point.

88. If a number is the opposite of another number, it is the negative of the other number.

89. The absolute value of a negative number is the additive inverse of the opposite of the number.

For each number, write **(a)** *the additive inverse and* **(b)** *the absolute value.*

90. $\dfrac{4}{3}$

91. $-\sqrt{5}$

92. $-3 + 3$

Evaluate each expression.

93. $-\left|-8\right|$

95. $\left|6\right| + \left|-10\right|$

94. $-\left|0\right|$

96. $\left|-5\right| + \left|-14\right| - \left|-2\right|$

Use inequality symbols to write each statement.

97. -8 is greater than -14.

98. -2 is less than 2.

99. a is between -4 and 5.

100. -11 is less than or equal to 15.

101. 6 is greater than or equal to 6.

102. t is between -3 and -1, including -3 and excluding -1.

103. w is between 0 and 11, inclusive.

1.2   Operations on Real Numbers

Objective 1   Add real numbers.

Find each sum.

1. $-4 + (-8)$

2. $-5.1 + (-7.3)$

3. $-7 + 5$

4. $3 + (-3)$

5. $5 + (-4)$

6. $-3 + (-4)$

7. $2 + (-7)$

8. $-4 + 8$

9. $-\dfrac{1}{2} + \dfrac{3}{8}$

10. $\dfrac{2}{7} + \left(-\dfrac{1}{2}\right)$

11. $-\dfrac{3}{5} + \left(-\dfrac{2}{5}\right)$

12. $-\dfrac{4}{5} + \dfrac{2}{3}$

13. $\dfrac{2}{11} + \left(-\dfrac{2}{3}\right)$

14. $\dfrac{5}{6} + \left(-\dfrac{5}{8}\right)$

15. $\left(-\dfrac{7}{12}\right) + \left(-\dfrac{3}{4}\right)$

16. $-\dfrac{4}{11} + \left(-\dfrac{7}{22}\right)$

17. $-16.32 + 2.27$

18. $-11.12 + 3.16$

19. $3 + 12 + (-6)$

20. $(-4) + (-2) + 6$

21. $(-12) + 7 + (-7)$

22. $(-11) + 5 + (-6)$

23. $3 + (-7) + (-8)$

24. $(-4) + (-5) + (-7)$

25. $|-13| + (-4) + 5$

26. $-5 + |-7| + 3$

Objective 2   Subtract real numbers.

Find each difference.

27. $3 - 7$

28. $7 - 11$

29. $-3 - 7$

30. $-3 - (-4)$

31. $7 - (-3)$

32. $-6.25 - (-2.47)$

33. $-12.86 - (-9.25)$

34. $\dfrac{2}{3} - \left(-\dfrac{1}{3}\right)$

35. $\dfrac{3}{5} - \left(-\dfrac{1}{3}\right)$

36. $3 + (-4) - 5$

37. $.4 - .3 - .7$

38. $-4 - 5 - (-3)$

39. $|-4| - 5 - (-2)$

40. $(-4) - (-3) - (-1)$

41 $12 - 3 - 7$

42. $3 - 1.2 - 7$

43. $|-20| - |9| - (-3)$

44. $3 - 5 - 8$

Objective 3   Multiply real numbers.

Find each product.

45. $(-4)(7)$

46. $(-5)(-3)$

47. $4(-.3)$

48. $6(-9)$

49. $(-1.2)(.7)$

50. $(-7)(3)(-5)$

51. $\left(\dfrac{3}{4}\right)\left(-\dfrac{9}{7}\right)$

52. $\left(-\dfrac{3}{8}\right)\left(-\dfrac{4}{5}\right)$

53. $\left(-\dfrac{3}{5}\right)(7)$

54. $3\left(-\dfrac{4}{21}\right)$

55. $12\left(-\dfrac{3}{10}\right)$

56. $\left(-\dfrac{4}{5}\right)\left(-\dfrac{15}{8}\right)$

57. $\dfrac{13}{11}\left(-\dfrac{33}{26}\right)$

58. $-1.2(-3.27)$

59. $2.4(-3.14)$

Objective 4   Find the reciprocal of a number.

Give the reciprocal of each number.

60. 13

61. -5

62. 7

63. -3

64. $.\overline{3}$

65. $-\dfrac{3}{5}$

66. $\dfrac{8}{7}$

67. $-\dfrac{11}{18}$

68. $\dfrac{23}{12}$

69. $\dfrac{5}{6}$

70. $.02$

71. $.35$

Objective 5   Divide real numbers.

Divide where possible.

72. $\dfrac{-3}{12}$

73. $\dfrac{15}{-30}$

74. $\dfrac{0}{-5}$

75. $\dfrac{-12}{14}$

76. $\dfrac{-5}{-20}$

77. $\dfrac{-7}{14}$

78. $\dfrac{-18.96}{2.4}$

79. $\dfrac{-10}{-.1}$

80. $\dfrac{16}{-.02}$

82. $\dfrac{-7}{0}$

84. $\dfrac{-3}{2} \div \left(-\dfrac{5}{7}\right)$

86. $\dfrac{\frac{7}{12}}{\frac{14}{3}}$

81. $\dfrac{0}{-7}$

83. $-\dfrac{2}{3} \div \dfrac{2}{9}$

85. $\dfrac{1}{2} \div (-2)$

1.2 Mixed Exercises

Decide whether each statement is **always true, sometimes true,** *or* **never true.** *If it is sometimes true, give an example where it is true and one where it is false.*

87. The sum of two numbers with unlike signs is positive.

88. The difference between two numbers with unlike signs is negative.

89. The sign of the product of two numbers with like signs is the same as the sign of the two numbers.

90. The quotient of two numbers with like signs is positive.

91. The quotient of 0 and a positive number is positive.

Perform the indicated operations.

92. $\dfrac{2}{3} + \left(-\dfrac{5}{3}\right) + \dfrac{7}{3}$

93. $-5 + |-4| + |4|$

94. $\dfrac{2}{3} + \left(-\dfrac{4}{7}\right) + \dfrac{3}{5}$

95. $\dfrac{1}{2} + \left(-\dfrac{3}{4}\right) + \dfrac{3}{8}$

96. $3 - (-5) - (-4)$

97. $2 - (-4) + (-3)$

98. $-8 + 6 + (-6)$

99. $-19 - (-9) + 9$

100. $4 - (-5) - 12 - (-6)$

101. $-5 - 8 - (-6) + 8$

1.3   Exponents, Roots, and Order of Operations

Objective 1   Use exponents.

Write each expression without exponents.

1. 3^3

2. 2^2

3. -5^2

4. -3^5

5. $(-3)^5$

6. $(-2)^3$

7. $-(-4)^2$

8. $\left(\dfrac{3}{5}\right)^2$

9. $\left(-\dfrac{4}{3}\right)^3$

10. $\left(\dfrac{5}{8}\right)^3$

11. $(.1)^2$

12. $(-.5)^2$

Objective 2   Identify exponents and bases.

Identify the exponent and the base. Do not evaluate.

13. 3^4

14. 12^3

15. $(-5)^3$

16. x^5

17. $-5y^6$

18. $8x^2$

19. -5^7

20. 8.1^5

21. 12^0

22. $(-6)^2$

23. $(5r)^4$

24. $(7p)^5$

Objective 3   Find square roots.

Find each square root that is a real number.

25. $\sqrt{16}$

26. $\sqrt{121}$

27. $-\sqrt{289}$

28. $\sqrt{0}$

29. $\sqrt{\dfrac{81}{4}}$

30. $\sqrt{\dfrac{256}{16}}$

31. $-\sqrt{-121}$

32. $-\sqrt{9}$

33. $\sqrt{-169}$

34. $\sqrt{.49}$

35. $\sqrt{.81}$

36. $-\sqrt{1600}$

37. $-\sqrt{-\dfrac{100}{121}}$

38. $-\sqrt{\dfrac{4}{49}}$

39. $\sqrt{144}$

40. $-\sqrt{.01}$

41. $-\sqrt{.25}$

42. $-\sqrt{-.01}$

43. $-\sqrt{-.25}$

44. $\sqrt{10,000}$

45. $\sqrt{-81}$

Objective 4   **Use the order of operations.**

Simplify.

46. $-4[2 - (-3)]$

47. $3(-6) + 2(-3)$

48. $-3 \cdot \dfrac{7}{6} - (-3)$

49. $-5 - 4(-5) + 7^2$

50. $-5 \cdot 2 - (-7)(-3)$

51. $(-3 - 3)(-7 - 5) - 5^2$

52. $12 \div 4 \cdot 3 \div 2$

53. $\dfrac{-7 + (-8)}{-3}$

54. $\dfrac{3(-5 + 7)}{-4}$

55. $\dfrac{4(-3) + (-4)(-3)}{-5 + 6 + 1}$

56. $-5\sqrt{9} + 4\sqrt{5} + 8^2$

57. $4^3 \div 2^5 + 3\sqrt{36}$

58. $2^4 \div 8 - \sqrt{121}$

59. $\dfrac{\left(\frac{3}{4} \cdot 12\right) + 7}{9 - \frac{3}{4} \cdot 8}$

60. $-\dfrac{1}{2}[4(-2) + 5(-7) + (-4)(-2)]$

61. $\dfrac{-[2 - (-4 + 1) + 1]}{\sqrt{9}(-1 + 3)}$

Objective 5   **Evaluate expressions for given value of variables.**

Evaluate if $a = -1$, $b = 3$, and $c = -5$.

62. $7a + 5b$

63. $2b - 3c$

64. $-3(a - 2c)$

65. $4b(3a - c)$

66. $-2c(3b + c)$

67. $-3a^2 + c$

68. $2a^3 - b^2$

69. $-a^2 + b^2$

70. $-a^2 - c^3$

71. $c^2 + a^2 + b^2$

72. $\dfrac{a + 3b}{2c}$

73. $\dfrac{3c - a}{5b}$

74. $\dfrac{2a + 4b^2}{3c - 2a}$

75. $\dfrac{6c - b}{7a^3 - 4c}$

1.3   Mixed Exercises

Evaluate.

76. 6^4

77. $(-1.1)^3$

78. $\sqrt{1.96}$

79. -12^2

80. $-\sqrt{100}$

81. $-\left(\dfrac{5}{3}\right)^4$

Simplify.

82. $-4(-3)^3 - 5(8-4)$

83. $\dfrac{6(-3)+(-5)^2(-4)}{11-3^2}$

Evaluate if x = 25, y = –3, and z = –1.

84. $\dfrac{3y-\sqrt{x}}{7yz}$

85. $\dfrac{2x+9y}{z-y^3}$

86. $\dfrac{4\sqrt{x}-3yz^3}{xyz^2}$

1.4 Properties of Real Numbers

Objective 1 **Use the distributive property.**

Use the distributive property to rewrite each expression.

1. $4(x-3)$ **3.** $-3(z-7)$ **5.** $4k+9k$ **7.** $3y-7y$ **9.** $4x+3y$

2. $7(t-5)$ **4.** $-(x-6)$ **6.** $6x+7x$ **8.** $6y-15y$ **10.** $7a+3b$

Objective 2 **Use the inverse properties.**

Complete each statement.

11. $7+\underline{\quad}=0$

12. $\underline{\quad}+(-4)=0$

13. $-3+\underline{\quad}=0$

14. $\underline{\quad}+1.5=0$

15. $-1.3+1.3=\underline{\quad}$

16. $-\dfrac{2}{3}+\underline{\quad}=0$

17. $\dfrac{1}{7}+\underline{\quad}=0$

18. $\dfrac{1}{8}\cdot\underline{\quad}=1$

19. $-6\cdot\underline{\quad}=1$

20. $-\dfrac{3}{4}\left(-\dfrac{4}{3}\right)=\underline{\quad}$

21. $5\cdot\underline{\quad}=1$

22. $\dfrac{4}{5}\cdot\underline{\quad}=1$

Objective 3 **Use the identity properties.**

Complete each statement.

23. $7+0=\underline{\quad}$

24. $-6+0=\underline{\quad}$

25. $\dfrac{4}{3}+\underline{\quad}=\dfrac{4}{3}$

26. $-\dfrac{3}{5}+\underline{\quad}=-\dfrac{3}{5}$

27. $\underline{\quad}+0=-2.5$

28. $8\cdot1=\underline{\quad}$

29. $-12\cdot1=\underline{\quad}$

30. $-\dfrac{4}{3}\cdot\underline{\quad}=-\dfrac{4}{3}$

31. $\underline{\quad}\cdot1=\dfrac{2}{3}$

32. $-\dfrac{7}{3}\cdot1=\underline{\quad}$

Objective 4 **Use the commutative and associative properties.**

Identify the property that justifies each statement.

33. $4+3=3+4$

34. $4(x+8)=4(8+x)$

35. $(4+3)+7=4+(3+7)$

36. $9\cdot8=8\cdot9$

37. $(5\cdot3)\cdot6=5\cdot(3\cdot6)$

38. $2+(x+9)=(2+x)+9$

39. $-5(x+z)=-5(z+x)$

40. $(4x)z=4(xz)$

41. $(ab)c=a(bc)$

42. $xy=yx$

43. $3[x(-4)]=3[(-4)x]$

44. $t(1+4)=(1+4)t$

Objective 5 **Use the multiplication property of 0.**

Complete each statement.

45. $426 \cdot 0 =$ _____

46. $-217 \cdot 0 =$ _____

47. $\dfrac{4}{13} \cdot$ _____ $= 0$

48. $0 \cdot$ _____ $= 0$

49. $-\dfrac{12}{25} \cdot 0 =$ _____

50. $0 \cdot 0 =$ _____

51. $0(x + y) =$ _____

52. $(4xy) \cdot 0 =$ _____

53. $-18($ _____ $) = 0$

54. $x \cdot$ _____ $= 0$

1.4 Mixed Exercises

Simplify each expression by removing parentheses and combining terms. In Exercises 55–66, identify the property or properties used.

55. $5r + 7r$

56. $a - 3a$

57. $2k + (3 + 8k)$

58. $6(s - 2t) + 3s$

59. $4(m - 2n) + 7n$

60. $(5 - 5)q$

61. $8p - 3(4 - 3p)$

62. $5j + 2(j - 4)$

63. $3(-2x + 5) - 15$

64. $-\dfrac{1}{8}(-8n)$

65. $5(3 - 2p) + 3(5p + 1)$

66. $(4m + 8) - (2m + 8)$

67. $-(2w - 7) + 11 + 3(4w - 6) - 5w$

68. $2(x - 3) - 7 - 6(2x - 5) + 7x$

69. $8 - 2(4d - 1) + 3(d - 6) + 2d$

Chapter 2

LINEAR EQUATIONS AND APPLICATIONS

2.1   Linear Equations in One Variable

Objective 1   Decide whether a number is a solution of a linear equation.

Decide whether or not the number is a solution of the given equation.

1. $5k = 35; 7$

2. $r + 12 = 10; 2$

3. $6m = 9; \dfrac{3}{2}$

4. $-s + 6 = 8; -2$

5. $\dfrac{3}{4}y = -12; -16$

6. $13 - k = 11; -2$

7. $6 - \dfrac{1}{2}p = 8; -4$

8. $\dfrac{1}{3}q - \dfrac{5}{3} = \dfrac{7}{3}; 4$

9. $8 = x - 4; 12$

10. $-22 = 6k - 4; -3$

Objective 2   Solve linear equations using the addition and multiplication properties of equality.

Solve and check each equation.

11. $3x + 6 + 2x = 4x - 3$

12. $7r - 4r + 2 = r - 4$

13. $12a + 5 = 5a - 2$

14. $6z - 3z + 5 - 4 = 3z - 5 + 4$

15. $11x - 14x - 7 + 8 = 4x + 5 - 2$

16. $9r - 4r + 8r - 6 = 10r - 11 + 2r$

17. $9x - 4 + 6x - 1 = 12 - 2x$

18. $18 - 3y + 11 = 7y + 6y - 27 - 9y$

19. $7t - 11t - 6 + 15 = -t + 17$

20. $21w - 14 + 8 - 26w = 3w - 18$

Objective 3   Solve linear equations using the distributive property.

Solve and check each equation.

21. $2(x + 4) = 3x - 2$

22. $3(x - 2) = 7x + 2$

23. $2(z - 1) + 3(z - 4) = 5$

24. $5(2p + 1) - (p + 3) = 7$

25. $-11z - (5 - 6z) = -(6 - 3z) + 1$

26. $-9w - (4 + 3w) = -(2w - 1) - 5$

27. $4t - 3(4 - 2t) = 2(t - 3) + 6t + 2$

28. $-[p - (4p + 2)] = 3 + (4p + 7)$

29. $6 - 5(2 - 3r) + 7 = 4(r - 6) - 17$

30. $-11a - (6a - 4) = -(5 - 3a) + 1$

31. $k + 2(-k + 4) - 3(k + 5) = -4$

32. $-r + 9(r - 5) + 4(3r + 4) = r - 5$

33. $4m - 3(5 - 2m) = 6(m - 3) + 2m + 1$

34. $6x - (8 - x) = 9[-2 - (5 + 2x) - 12]$

Objective 4   **Solve linear equations with fractions or decimals.**

Solve and check each equation.

35. $\dfrac{1}{3}m = 17$

36. $-\dfrac{5}{7}y = 14$

37. $-\dfrac{4}{5}x = 8$

38. $\dfrac{z}{2} + \dfrac{z}{3} = 5$

39. $-\dfrac{4}{3}x + 5 = 7$

40. $-\dfrac{4}{7}q - 2 = 6$

41. $\dfrac{6w}{7} = 24$

42. $\dfrac{m}{6} + \dfrac{m}{4} = -9$

43. $\dfrac{p-2}{3} + \dfrac{p}{4} = \dfrac{1}{2}$

44. $\dfrac{y-8}{5} + \dfrac{y}{3} = -\dfrac{12}{5}$

45. $\dfrac{2x+5}{5} - \dfrac{3x+1}{2} = \dfrac{7-x}{2}$

46. $\dfrac{x-5}{2} - \dfrac{x+6}{3} = -4$

47. $.06x + .14(x + 500) = 130$

48. $.35(140) + .15w = .05(w + 1100)$

49. $.07(w - 14) - .05w = 3.22$

50. $.08(x - 9) = 1.38 - .02x$

51. $5.24 - .06(x + 9) = .04x$

52. $.04(x + 6) - .02x = 3.16$

Objective 5   **Identify conditional equations, contradictions, and identities.**

Decide whether each equation is a **conditional equation,** *an* **identity,** *or a* **contradiction.** *Give the solution set.*

53. $5m - 2(4 - m) = 6$

54. $2[3 - 4(5 - r)] = 2(3r - 11)$

55. $6t - 3(5 + 2t) = -12$

56. $7(2 - 5b) - 32 = 10b - 3(6 + 15b)$

57. $6(3 - 4x) + 10 = -15x + 3(2 - 3x)$

58. $2(2y - 5) - 3(4 - y) = 7y - 22$

59. $28 - 4k = 36 + 2(2k - 4)$

60. $7(3 - 4q) - 10(q - 2) = 19(5 - 2q)$

61. $6z - 3(8z - 4) = 2(4 - 7z) - 4(z - 1)$

62. $13p - 9(3 - 2p) = 3(10p - 9) + 1$

2.1   Mixed Exercises

Decide whether or not the given number is a solution of the equation.

63. $3y - 8y = 45;\ 5$

64. $15 - 7x = 12;\ \dfrac{3}{7}$

Solve and check each equation.

65. $5x - 7x + 14 = 4 - 11 - 9x$

66. $8(p - 4) = 6p - 10$

67. $9q + 114 - 2q = 15 - 6q + 8$

68. $6a - 3(5a + 2) = 4 - 5a$

69. $3(x - 6) - 4(8 - 3x) = 10(x - 4) + 4$

70. $5(1 - 2x) + 3(x - 4) = -7(1 + x)$

71. $\dfrac{2}{3}k = 8$

72. $-\dfrac{5}{7}y = 15$

73. $\dfrac{a + 8}{6} + \dfrac{4a}{9} = \dfrac{2a + 12}{9}$

74. $\dfrac{2a + 3}{5} - \dfrac{3a - 1}{2} = \dfrac{4a + 7}{2}$

75. $4(x - 12) - 8(x + 1) = 2(28 - 2x)$

76. $3(t - 4) + 5(6 - 2t) = 7(2 - t) + 4$

77. $.05(x - 2) + .06x = .78$

78. $.08x - .05(x - 94) - 4.88 = 0$

2.2   Formulas

Objective 1   Solve a formula for a specified variable.

Solve each formula for the specified variable.

1. $I = prt$ for r

2. $V = \dfrac{1}{3}Bh$ for B

3. $d = 2r$ for r

4. $A = LW$ for L

5. $P = 4s$ for s

6. $P = a + b + c$ for b

7. $V = LWH$ for H

8. $V = \pi r^2 h$ for h

9. $C = \dfrac{5}{9}(F - 32)$ for F

10. $F = \dfrac{9}{5}C + 32$ for C

Solve each equation for the specified variable.

11. $y = \dfrac{3xy + 7}{x + 2}$ for y

12. $r = \dfrac{x + 7}{x}$ for x

13. $\dfrac{2 + y}{3} = \dfrac{y}{x}$ for y

14. $xy - 3 = \dfrac{x}{2}$ for x

Objective 2   Solve applied problems using formulas.

Solve each problem using the appropriate formula.

15. If the distance is 60 kilometers and the rate is 20 kilometers per minute, find the time.

16. A cord of wood contains 128 cubic feet of wood. A stack of wood is 4 feet high and 8 feet long. How wide must it be to contain a cord?

17. Two sides of a triangle are 8 and 13. Find the third side if the perimeter is 37.

18. The area of a trapezoid is 60 square feet. If the bases are 6 feet and 14 feet, find the altitude of the trapezoid.

19. A woman travels 600 kilometers in 4 hours. Find the average rate.

20. Find the width of a rectangle if the perimeter is 20 inches and the length is 6 inches.

21. If $700 earns $112 simple interest in 2 years, find the rate of interest.

22. The Fahrenheit temperature is 104°. Find the Celsius temperature.

23. A circle has a circumference of 36π meters. Find the radius of the circle.

24. Base is 70; percentage is 14. Find the amount.

25. If $685 earns $274 simple interest in 4 years, find the rate of interest.

26. A cylindrical can has a surface area of 30π square inches. The radius of the can is 3 inches. Find the height of the can.

Objective 3   Solve percent problems.

Find the required percent or percentage.

27. An alcohol and water mixture measures 45 liters. The mixture contains 9 liters of alcohol. What percent of the mixture is alcohol?

28. There are 6 liters of sulfuric acid in a 24-liter mixture of sulfuric acid and water. What percent of the mixture is sulfuric acid?

29. An intermediate algebra class has an enrollment of 48 students of which 30 are female. What percent of the class is male?

30. At a particular college 120 students are enrolled in organic chemistry. Of this group 18 are freshman, 52 are sophomores, 48 are juniors, and 2 are seniors. What percent of the organic chemistry students is made up of juniors?

31. A salesperson earns 15% commission each month on all that she sells. In a particular month she sold $12,000 in goods. What was her commission for that month?

32. At a large university, 40% of the student body is from out of state. If the total enrollment at the university is 25,000 students, how many are from out of state?

33. A bank account paid $264.50 last year on a deposit of $11,500. What is the rate of interest?

34. The purchase price of a new car is $12,500. In order to finance the car a purchaser is required to make a minimum down payment of 20% of the purchase price. What is the minimum down payment required?

35. In a class of freshmen and sophomores only, there are 85 students. If 34 are freshmen, what percent of the class are sophomores?

36. Twelve percent of a college student body has a grade point average of 3.0 or better. If there are 1250 students enrolled in the college, how many have a grade point average of less than 3.0?

37. A certificate of deposit pays $117 simple interest in one year on a principal of $4500. What interest rate is being paid on this deposit?

38. Mark Johnson invested $2500 in stock one year ago. During the year the stock increased in value by $162.50. What interest rate has Mark's investment earned?

39. A salesperson earned $33,250 on annual sales of $950,000. What is her rate of commission?

40. The sale price of a leather coat is $520. This represents 20% off the regular price. What is the regular price?

41. The sale price of a piece of furniture is $1020. This represents 15% off the regular price. What is the regular price?

2.2 Mixed Exercises

Solve each formula for the specified variable.

42. $A + B + C = 180°$ for C

43. $C = \pi d$ for d

Solve each problem.

44. Maria travels 80 kilometers at a rate of 20 kilometers per hour. What is her travel time?

45. Vinegar and water are mixed to give 30 liters of liquid. The mixture contains 5 liters of vinegar. What is the percent of vinegar in the mixture?

46. In 32 ounces of fruit drink, 25% is pure fruit juice. How many ounces of pure fruit juice are in the mixture?

47. Tyler travels 260 miles in 5 hours. What is his rate?

48. A triangle has sides of 3.5 meters, 4.2 meters, and 2.7 meters. What is the perimeter?

49. The circumference of a circle is 380π feet. Find the diameter of the circle.

50. The Fahrenheit temperature is 41°. Find the Celsius temperature.

51. The purchase of a home requires a 20% down payment. If the purchase price is $140,000, how much is the down payment?

52. The sale price of a stereo system is $420. This represents 30% off the regular price. What is the regular price?

2.3   Applications of Linear Equations

Objective 1   Translate from words to mathematical expressions.

Translate each verbal phrase into a mathematical expression. Use x to represent the unknown number.

1. A number increased by 13

2. A number decreased by –5

3. The product of a number and 7

4. 8 less than a number

5. 10 more than a number

6. 14 decreased by twice a number

7. The ratio of a number and 5

8. The quotient of a number and –3

9. –7 increased by 4 times a number

10. The quotient of 4 more than a number and 9

Objective 2   Write equations from given information.

Use the variable x for the unknown, and write an equation representing the verbal sentence. Do not solve.

11. The sum of a number and 9 is 45.

12. If twice a number is decreased by 3 the result is 25.

13. The product of a number and 6 is 7 plus the product of 5 times the number.

14. The quotient of a number and 7 is 6.

15. The sum of a number and twice the number is 12.

16. The ratio of a number and the difference between the number and 3 is 17.

17. Twice a number is 3 times the sum of the number and 7.

18. 27 is 4 more than 7 times a number.

19. 40% of the sum of a number and 3 is 5.

20. 18 minus a number is equal to the number times 4.

Objective 3   Distinguish between expressions and equations.

Decide whether each is an expression or an equation.

21. $3(z - 9) + 8(2z - 4)$

22. $3(z - 9) = 8(2z - 4)$

23. $-y + 4 - 2(y - 8)$

24. $-y + 4 = -2(y - 8)$

25. $\dfrac{k}{3} - \dfrac{k+5}{4} = 12$

26. $\dfrac{k}{3} - \dfrac{k+5}{4} - 12$

Objective 4   **Use the six steps in solving an applied problem.**

Solve each problem.

27. The length of a rectangle is 12 feet less than twice the width. The perimeter is 60 feet. Find the length and width.

28. The width of a rectangle is $\frac{1}{4}$ of the difference between the length and 1 meter. If the perimeter of the rectangle is 62 meters, find the length and width.

29. In a triangle, the shortest side is 3 meters less than the longest side. The length of the third side is 11 meters less than twice the length of the longest side. If the perimeter is 22 meters, find the lengths of the three sides of the triangle.

30. In a triangle with two sides of equal length, the third side is 6 feet less than the sum of the lengths of the two equal sides. The perimeter of the triangle is 26 feet. Find the lengths of the three sides of the triangle.

31. The perimeter of a triangle with three equal sides is 26 meters longer than one side. What are the lengths of a side and the perimeter?

32. Mrs. Henry has 36 feet of fencing, which she plans to install to form a square yard for her dog. If the back of her house becomes one side of the square with the other sides requiring fencing, how long should each side be?

33. John Davis has purchased a rare print which he plans to frame. The print is a rectangle 86 centimeters by 54 centimeters. If John uses a framing material that is 4 centimeters wide to frame the print, what will the perimeter of the framed print be?

34. Judy needs a rectangular storage area. She wants the length to be 3 feet more than the width, and the perimeter can only be 22 feet. Find the length and width of the storage area.

35. A lake manager wants to stock a small lake with 11,550 fish, some bass and some walleyes. He wants to use 1830 more bass. How many of each kind of fish does he need?

36. In a student government election, two candidates received 3215 votes. The losing candidate received 637 fewer votes than the winning candidate. How many votes did each candidate receive?

Objective 5   **Solve percent problems.**

Solve each percent problem.

37. If 18% of a certain number is 63, find the number.

38. The result of increasing a certain number by 20% of the number is 54. Find the number.

39. The result of decreasing a certain number by 16% of the number is 546. Find the number.

40. The sale price of a particular item that had been reduced by 20% is $28. What was the original price?

41. A savings account earns interest at an annual rate of 4%. At the end of one year Sheri Minkner would like to have a total of $5200 in the account. How much should she deposit now?

42. A merchant in a small store took in $508.80 in one day. This included merchandise sales and the 6% sales tax. Find the amount of merchandise sold.

43. The original price of a television set was $350. On sale the set sold for $245. By what percent was the set reduced?

44. Last year Mark and Paula spent, on average, $120 per week for groceries. This year they spend $126 per week on average. By what percent has their weekly expenditure for groceries increased?

45. A certificate of deposit earns $2\frac{3}{4}\%$ annually. If $6200 is invested, what is the certificate of deposit worth in one year?

46. A certificate of deposit is worth $657.60 after a year. If the annual rate is $2\frac{3}{4}\%$, what was the certificate worth originally?

Objective 6   **Solve investment problems.**

Solve each investment problem.

47. Tracy Sudak invested some money at 5% and $300 more than twice this amount at 7%. Her total annual income from the two investments is $325. How much is invested at each rate?

48. Larry Frank invested some money at 8% simple interest and $700 less than this amount at 7%. His total annual income from the interest was $584. How much was invested at each rate?

49. Gena Zarcone invested some money at 8% and $500 more than that amount at 6%. Her total annual interest was $58. How much did she invest at each rate?

50. Desiree Phillips invested some money at 9% and $100 less than three times that amount at 7%. Her total annual interest was $83. How much did she invest at each rate?

51. Louisa Hernandez invested $16,000 in bonds paying 7% simple interest. How much additional money should she invest at 4% simple interest so that the average return on the two investments is 6%?

52. Zach Cain invested some money at 5.5% and $2000 less than twice this amount at 4%. His total annual income from the interest was $257.50. How much was invested at each rate?

53. Courtney Taylor placed $24,000 in an account paying 6.5%. How much additional money should she deposit at 4% so that the average return on the two investments is 6%?

54. Jacob Hughes has $48,000 invested in stocks paying 6%. How much additional money should he invest in certificates of deposit paying 2.5% so that the average return on the two investments is 4%.

Objective 7   **Solve mixture problems.**

Solve each mixture problem.

55. How many liters of 20% alcohol solution must be mixed with 10 liters of a 50% solution to get a 30% solution?

56. How many liters of pure alcohol must be mixed with 20 liters of an 18% alcohol solution to obtain a 20% alcohol solution?

57. How many milliliters of water must be mixed with 25 milliliters of a 32% sulfuric acid solution to obtain a 20% sulfuric acid solution?

58. How many gallons of a 20% alcohol solution must be mixed with 15 gallons of a 12% alcohol solution to obtain a 14% alcohol solution?

59. How many pints of a 6% disinfectant solution must be mixed with 12 pints of a 10% disinfectant solution to obtain a 7% disinfectant solution?

60. How many pounds of candy worth $7 per pound must be mixed with candy worth $4.50 per pound to make 100 pounds of candy worth $6 per pound?

61. How many ounces of a 35% solution of acid must be mixed with a 60% solution to get 20 ounces of a 50% solution?

2.3   Mixed Exercises

Use the variable x for the unknown, and write an equation representing the verbal sentence. Do not solve.

62. If twice a number is added to 50, the result is the number decreased by 6.

63. The quotient of a number and 4, added to twice the number, is 8.

64. A number multiplied by 30 is 87 more than the number.

Solve each problem.

65. The number of students enrolled in environmental studies at Middleton College in 1970 was 450. In 1995, this number rose to 675. What was the percent increase?

66. The length of a rectangle is 7 inches longer than twice the width. What are the dimensions of the rectangle if its perimeter is 74 inches?

67. How many gallons of a 5% disinfectant solution must be mixed with 5 gallons of a 10% solution to obtain a 7% solution?

68. Jim Monroe wishes to make $235,000 selling his home. If his realtor must earn a 6% commission, for what price should Jim sell his home?

69. The rate of inflation rose by 14% from 1990 to 1994. If Sylvia Woodard earned $22,000 in 1990, what should her earnings have been in 1994 to keep up with inflation?

70. Rajesh Sheh invested $4000 at 4.5%. How much should he invest at 7% to have an average return on the two investments of 6%?

71. How many pounds of peanuts worth $3 per pound must be mixed with mixed nuts worth $5.50 per pound to make 40 pounds of a mixture worth $5 per pound?

72. Andria Pihos invested some money at 6% and $100 less than that amount at 9%. Her total annual interest was $96. How much did she invest at each rate?

2.4   Further Applications of Linear Equations

Objective 1   **Solve problems about different denominations of money.**

Solve each problem.

1. A collection of dimes and nickels has a total value of $2.70. The number of nickels is 2 more than twice the number of dimes. How many of each type of coin are in the collection?

2. Erica Clayton's piggy bank has quarters and nickels in it. The total number of coins in the piggy bank is 50. Their total value is $8.90. How many of each type are in the piggy bank?

3. At the end of each day, Vanessa Gillis throws all the change from her purse into a box. The box contains only pennies, nickels, and dimes. At the end of the week she found that the total value of the coins was $4.80. If the number of dimes was 1 more than the number of nickels and the number of pennies was 6 more than the number of nickels, how many of each type of coin was in the box?

4. A school play was attended by 450 people. Adult admission to the play was $3 while student admission was $2. If the total receipts for the play were $1180, how many students attended?

5. James Bach purchased some stamps for $53. He purchased three kinds of stamps in denominations of $.19, $.29, and $2.90. The number of $.19 stamps was the same as the number of $.29 stamps, which was 5 times the number of $2.90 stamps. How many of each kind of stamp did he purchase?

6. Anthony Silva sells two sizes of jars of Tan Joy peanut butter. The large size sells for $2.60 and the small size sells for $1.80. He has 80 jars worth $164. How many of each size jar does he have?

Objective 2   **Solve problems about uniform motion.**

Solve each problem.

7. Laura Dawson can get to school in $\frac{1}{4}$ hour if she rides her bike. It takes her $\frac{3}{4}$ hour if she walks. Her speed when walking is 6 miles per hour slower than her speed when riding. What is her speed when riding?

8. Two cars leave a city at the same time, but in opposite directions. One travels at 56 miles per hour, and the other travels at 60 miles per hour. In how many hours will they be 522 miles apart?

9. Two boats leave a port at the same time, one traveling north and the other south. The one traveling north steams at 13 miles per hour, and the one traveling south steams at 19 miles per hour. In how many hours will they be 232 miles apart?

10. On a 690-mile automobile journey, Sanying Jin averaged 55 miles per hour on the first part of the trip and 60 miles per hour on the second part of the trip. How long did the entire journey take if the two parts each took the same number of hours?

11. Leslie Kramer's drive to work takes 1 hour. If she increases her average speed by 10 miles per hour, the trip takes $\frac{3}{4}$ hour. Find the distance she drives to work.

12. A 280-mile automobile trip took Meyer Kaufmann a total of 5 hours. His average speed for the last 3 hours was 10 miles per hour faster than his average speed for the first 2 hours. Find his average speed for the last 3 hours of the trip.

13. Hans and Greta live 21 miles apart. At 1:00 P.M. they start riding their bicycles toward each other and meet at 1:45 P.M. If Greta's average speed is 4 miles per hour faster than Hans's average speed, find Greta's average speed.

14. Two trains leave at the same time from cities that are 450 miles apart, and travel toward each other. One train travels 20 miles per hour slower than the other. They pass each other after 3 hours. Find the rate of each train.

15. Two distance runners leave from the same point, traveling in opposite directions. One runs 2 miles per hour slower than the other. After 2 hours, they are 36 miles apart. Find the rate of each runner.

16. Two airplanes leave Cleveland Hopkins airport at the same time, traveling in opposite directions. One travels 25 miles per hour faster than the other. After 2 hours they are 490 miles apart. Find the rate of each airplane.

Objective 3   **Solve problems involving the angles of a triangle.**

Find the measure of each angle in the triangles shown.

17.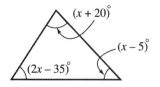

18.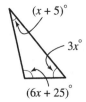

19. One angle of a triangle measures 30° larger than a second angle. The third angle measures 4 times the second angle. Find the measure of each angle.

20. One angle of a triangle measures twice the second angle. The measure of the third angle equals the sum of the measures of the first two angles. Find the measure of each angle.

21. The measure of an angle is 15° larger than 10 times its supplement. Find the measure of each angle.

22. The measure of an angle is 10° smaller than 4 times the measure of its supplement. Find the measure of each angle.

23. The measure of an angle is 10° smaller than 4 times the measure of its complement. Find the measure of each angle.

2.4   Mixed Exercises

Solve each problem.

24. Yolanda Lamar has 30 coins in her change purse, consisting of pennies and nickels. The total value of the money is $.94. How many of each type of coin does she have?

25. Two planes leave O'Hare Airport in Chicago at the same time. One travels east at 550 miles per hour, and the other travels west at 500 miles per hour. How long will it take for the planes to be 2100 miles apart?

26. The measure of one angle of a triangle is 30° more than that of a second angle. The measure of the third angle is half the measure of the second angle. Find the measure of each angle.

27. Two steamers leave ports on a river at the same time, traveling toward each other. The ports are 66 miles apart. If each is traveling at 12 miles per hour, how long will it take them to pass each other?

28. On one day, 660 people visited the local zoo. Senior citizens attended free, children paid $3 each, and adults paid $5 each. If twice as many children visited as senior citizens, and $2292 was collected, how many of each age group visited the zoo that day?

29. Emily Rausch has 28 coins in her pocket, consisting of nickels and dimes. The total value of the money is $2.70. How many of each type of coin does she have?

30. An angle measures 18° more than twice its supplement. Find the measure of each angle.

31. Jane leaves Nashville to visit her cousin Linda in Napa, 80 miles away. She travels at an average speed of 50 miles per hour. One-half hour later Linda leaves to visit Jane, traveling at an average speed of 60 miles per hour. How long after Linda leaves will it be before they meet?

32. Two cars leave from towns that are 180 miles apart at the same time, traveling toward each other. One car travels 10 miles per hour faster than the other. They meet each other after 2 hours. Find the rate of each car.

33. In a run for charity Ted runs at a speed of 5 miles per hour. Bob leaves 10 minutes after Ted and runs at 6 miles per hour. How long will it take Bob to catch up with Ted? (Hint: Change minutes to hours.)

Chapter 3

LINEAR EQUATIONS AND INEQUALITIES

3.1 Linear Inequalities in One Variable

Objective 1 Graph intervals on a number line.

Write each inequality in interval notation and graph it.

1. $x > -4$
2. $x < 6$
3. $x \geq 2$
4. $x \leq 0$
5. $-2 \leq x \leq 3$
6. $1 < x \leq 5$
7. $-4 \leq x < 0$
8. $-5 < x < -1$

Objective 2 Solve linear inequalities using the addition property.

Solve each inequality, giving its solution set in interval form.

9. $x + 7 \leq 4$
10. $p - 3 > 6$
11. $x + 5 > -1$
12. $t - 2 \geq 7$
13. $z + 8 < 8$
14. $6s < 7s - 3$
15. $8s < 7s - 3$
16. $x - 3 \geq -5$
17. $a + 5 > 6$
18. $y - 7 \leq 8$
19. $p + 2 > 6$
20. $6 < -3t + 4t$

Objective 3 Solve linear inequalities using the multiplication property.

Solve each inequality, giving its solution set in both interval and graph forms. Check your answers.

21. $2r \leq 6$
22. $-3a < 9$
23. $2z < -8$
24. $-2x \geq 5$
25. $-16z \leq -64$
26. $-\dfrac{1}{2}k \geq 5$
27. $-\dfrac{3}{4}r \geq 27$
28. $-\dfrac{3}{5}x \leq -6$
29. $-\dfrac{2}{7}x > -4$
30. $20 - 3(2p + 4) \leq -10p$
31. $14 - 3(p + 2) < -5p$
32. $8 - 4(p - 2) > -6p$
33. $12 - 2(p - 3) \geq -8p$

Objective 4 Solve linear inequalities with three parts.

Solve each inequality, giving its solution set in both interval and graph forms. Check your answers.

34. $-2 < y - 3 < 6$
35. $-6 < k + 2 < 8$
36. $10 < z + 5 < 14$
37. $-4 \leq a + 5 < -2$
38. $-4 < 6 - 2x < -2$
39. $4 > 3a + 4 > -4$
40. $-3 \leq \dfrac{2t + 1}{6} \leq 5$
41. $-3 \leq \dfrac{6q - 1}{4} \leq 0$
42. $-5 \leq \dfrac{3}{4}x + 1 \leq 10$

43. $-4 \le \dfrac{2}{5}x + 2 \le 6$

Objective 5   **Solve applied problems using linear inequalities.**

Find the unknown number. Give your answer as a verbal statement.

44. 3 times a number is between –6 and 6.

45. Half a number is between –5 and 3.

46. When 2 is added to twice a number, the result is less than 8.

47. If 6 is subtracted from a number, the result is at least 4.

48. Half a number is subtracted from 7, giving a result of at least 3.

49. When 3 times a number is subtracted from 8, the result is greater than or equal to 5.

50. Six times a number minus 7 is less than zero.

Answer the question.

51. Which symbol is used to represent "is at most" in an inequality?

52. Which symbol is used to represent "is no less than" in an inequality.

53. Which symbol is used to represent "is at least" in an inequality.

54. Which symbol is used to represent "no more than" in an inequality.

Solve each problem.

55. Megan has 3 times as many nickels as dimes and she has at least 20 coins. At least how many dimes does she have?

56. Andrew must have an average of 80% of the points on four exams to receive a B in the class. He has earned 78%, 83%, and 75% on the first three exams. What is the lowest score he can earn on a 100-point test to guarantee a B in the class?

57. The college of your choice charges $8200 tuition annually. If you can make no more than $2050 at your summer job, how many summers will you have to work to earn at least enough for one year's tuition?

3.1   Mixed Exercises

Solve each inequality, giving its solution set in both interval and graph forms.

58. $-\dfrac{2}{5}x < \dfrac{3}{5}$

59. $-4 \le \dfrac{1-2x}{6} \le 0$

60. $4 < 3m + 5 < 7$

61. $-3 < 2t + 4 < 6$

62. $-3 \le \dfrac{3y-1}{-4} \le 1$

63. $-\dfrac{4}{7}y \ge \dfrac{3}{14}$

Solve each problem.

64. Margaret Driscoll gets scores of 88 and 78 on her first two tests. What score must she make on her third test to keep an average of 80 or greater?

65. A Christmas tree with only red and green lights has twice as many green lights as red lights. If it has at least 15 lights, at least how many red lights does it have?

66. A nurse must make sure that a patient receives at least 30 units of a certain drug each day. This drug comes from red pills or yellow pills, each of which provides 3 units of the drug. The patient must have twice as many red pills as yellow pills. At least how many yellow pills will satisfy the requirement?

67. To make a profit, a potter's sales of pots must be greater than his costs to make the pots. The potter sells pots at a price of $8 that cost him $5.50 to make. He also has a basic cost of $15 on each batch of pots. How many pots must he sell in a batch in order to make a profit?

3.2   Set Operations and Compound Inequalities

Objective 1   **Find the intersection of two sets.**

Find each intersection of sets.

1. $\{0, 1, 2, 3\} \cap \{2, 3, 4, 5\}$

3. $\{7, 8, 9, 10\} \cap \emptyset$

2. $\{-6, -5, -4\} \cap \{-3, -2, -1\}$

4. $\{2, 6, 8\} \cap \{2, 6, 8\}$

Let $A = \{0, 1, 2, 3, 4, 5\}$, $B = \{2, 4, 6, 8, 10\}$, $C = \{1, 3, 5, 7, 9\}$, $D = \{0, 2, 4\}$, and $E = \{0\}$. Specify each intersection.

5. $A \cap B$

7. $A \cap D$

9. $B \cap C$

6. $A \cap C$

8. $A \cap E$

10. $B \cap D$

Objective 2   **Solve compound inequalities with the word and.**

For each compound inequality, give the solution set in both interval and graph forms.

11. $r < 3$ and $r > 0$

16. $2z + 1 < 3$ and $3z - 3 > 3$

12. $m \leq 4$ and $m \leq 7$

17. $3x + 2 < 11$ and $2 - 3x \leq 14$

13. $t \geq -2$ and $t \geq 1$

18. $5t > 0$ and $5t + 4 \leq 9$

14. $x - 3 \leq 6$ and $x + 2 \geq 7$

19. $q < -1$ and $q \geq 2$

15. $2q < -2$ and $q + 3 > 1$

20. $r \geq 2$ and $r \leq -2$

Objective 3   **Find the union of two sets.**

Find each union of sets.

21. $\{0, 1, 2, 3\} \cup \{2, 3, 4, 5\}$

23. $\{7, 8, 9, 10\} \cup \emptyset$

22. $\{-6, -5, -4\} \cup \{-3, -2, -1\}$

24. $\{2, 6, 8\} \cup \{2, 6, 8\}$

Let $A = \{1, 2, 3, 4, 5, 6\}$, $B = \{0, 2, 4, 6, 8, 10\}$, $C = \{1, 3, 5, 7, 9\}$, $D = \{1, 2, 3\}$, and $E = \{0\}$. Specify each union.

25. $A \cup B$

27. $A \cup D$

29. $B \cup C$

26. $A \cup C$

28. $A \cup E$

30. $B \cup D$

Objective 4   **Solve compound inequalities with the word or.**

For each compound inequality, give the solution set in both interval and graph forms.

31. $m > 4$ or $m < -1$

34. $k \leq 3$ or $k \geq 6$

32. $y \leq 1$ or $y \geq 6$

35. $r \geq -1$ or $r \geq 4$

33. $a \leq 2$ or $a \geq 6$

36. $p \geq -1$ or $p \leq 6$

37. $q + 3 > 7$ or $q + 1 \leq -3$

38. $s - 5 > 0$ or $s + 7 < 6$

39. $4x < x - 5$ or $6x > 2x + 3$

40. $3 > 4m + 2$ or $7m - 3 \geq -2$

3.2 Mixed Exercises

Let $A = \{0\}$*,* $B = \{2, 4\}$*,* $C = \{1, 3, 5, 9\}$*, and* $D = \{0, 1, 2, 3, 4, 5\}$*. Specify each set.*

41. $A \cup B$ **42.** $C \cap D$ **43.** $A \cup D$ **44.** $B \cap C$

For each compound inequality, give the solution set in both interval and graph forms.

45. $z \geq 0$ or $z \leq -2$

46. $y \geq -2$ and $y < 3$

47. $2r + 4 \geq 8$ or $4r - 3 < 1$

48. $4t < 2t + 10$ or $t - 3 > 3$

49. $z \leq 2$ and $z \geq 6$

50. $x \geq -4$ or $x \leq 5$

51. $-2x + 1 < 3$ and $3x \leq 12$

52. $7y + 5 \leq 3$ and $-3y \geq -9$

3.3   Absolute Value Equations and Inequalities

Objective 1   Use the distance definition of absolute value.

Graph the solution set of each equation or inequality.

1. $|m| = 7$

2. $|q| = 0$

3. $|k| < 8$

4. $|r| > 2$

5. $|x| \geq 6$

6. $|y| \geq 0$

7. $|p| \geq -2$

8. $|p| < 3$

9. $|x| \leq 10$

10. $|t| \leq 0$

Objective 2   Solve equations of the form $|ax+b| = k$, for $k > 0$.

Solve each equation.

11. $|t-4| = 7$

12. $|m+4| = 8$

13. $|3-q| = 7$

14. $|3k-1| = 6$

15. $|5-t| = 3$

16. $|m+6| = 2$

17. $|2x+3| = 10$

18. $|2r+3| = 0$

19. $|5r-15| = 0$

20. $\left|5 - \frac{4}{3}x\right| = 9$

Objective 3   Solve inequalities of the form $|ax+b| < k$ and of the form $|ax+b| > k$, for $k > 0$.

Solve each inequality and graph the solution set.

21. $|x-2| > 8$

22. $|q+5| > 15$

23. $|n+5| < 8$

24. $|5-z| < 8$

25. $|5r+2| < 18$

26. $|2r-9| \geq 23$

27. $|3q-5| + 2 \geq 6$

28. $|-k-3| \geq 1$

29. $|2z+4| + 6 > 8$

30. $|p-5| - 5 \geq 0$

31. $|2-z| \leq 3$

32. $|4y-1| - 3 \leq -1$

Objective 4 Solve absolute value equations that involve rewriting.

Solve each equation.

33. $|a| + 5 = 7$

34. $|z| - 6 = 3$

35. $|s| + 2 = 0$

36. $|y| - 5 = -7$

37. $|5 + y| + 3 = 7$

38. $|7t + 5| + 6 = 14$

39. $|5 - 2w| + 7 = 5$

40. $|2w - 1| + 7 = 12$

41. $\left| 2 - \dfrac{1}{2}x \right| - 5 = 18$

42. $|4t + 3| + 8 = 10$

Objective 5 Solve equations of the form $|ax + b| = |cx + d|$.

Solve each equation.

43. $|2x + 6| = |3x - 9|$

44. $|a - 4| = |a - 3|$

45. $|y + 3| = |2y - 5|$

46. $|5 - z| = |2z + 3|$

47. $|3 - a| = |a + 5|$

48. $|y + 5| = |3y + 1|$

49. $|2x - 8| = |6x + 7|$

50. $|2p - 4| = |7 - p|$

51. $\left| p - \dfrac{1}{2} \right| = \left| \dfrac{1}{2}p - 1 \right|$

52. $\left| y - \dfrac{1}{4} \right| = \left| \dfrac{1}{2}y + 1 \right|$

Objective 6 Solve special cases of absolute value equations and inequalities.

Solve each equation.

53. $|2x - 4| = -6$

54. $\left| 7 + \dfrac{1}{2}x \right| = 0$

55. $|p| = 0$

56. $|a| \leq -2$

57. $|k + 5| \leq -2$

58. $|3 - 2x| + 5 \leq 1$

59. $|4 + t| < 0$

60. $|2 + q| < 0$

61. $|m - 2| \geq -1$

62. $|3p + 4| > -7$

3.3   Mixed Exercises

Solve each equation.

63. $\left|\,2y+5\,\right|=3$

64. $\left|\dfrac{1}{2}x-3\right|=4$

65. $\left|\,x+7\,\right|=-3$

66. $\left|\,r+5\,\right|=6$

67. $\left|\,r\,\right|+7=5$

68. $\left|\,3t+2\,\right|-9=18$

69. $\left|\,3x-2\,\right|=\left|\,5x+8\,\right|$

70. $\left|\dfrac{2}{3}z+1\right|=\left|\dfrac{1}{3}z-1\right|$

Solve each inequality and graph the solution set.

71. $\left|\,3-2a\,\right|<7$

72. $\left|\,3r-9\,\right|\le 10$

73. $\left|\,2x+7\,\right|>9$

74. $\left|\,3-y\,\right|\ge 7$

75. $\left|\,2z+1\,\right|+4\ge 15$

76. $\left|\,4x\,\right|+3>5$

77. $\left|\,3y-1\,\right|+3\le 5$

78. $\left|\,4n-3\,\right|+6<11$

Chapter 4

GRAPHS, LINEAR EQUATIONS, AND FUNCTIONS

4.1 The Rectangular Coordinate System

Objective 1 Plot ordered pairs.

Locate each point on the rectangular coordinate system.

1. $(2, 1)$ **3.** $(6, -1)$ **5.** $(4, 0)$ **7.** $(0, -4)$

2. $(-3, 5)$ **4.** $(-5, -4)$ **6.** $(-3, 0)$ **8.** $(0, 6)$

Objective 2 Find ordered pairs that satisfy a given equation.

Complete the ordered pairs for the given equation.

	Equation	*Ordered Pairs*
9.	$x + y = 6$	$(0, \ ), (\ , 0), (2, \ ), (\ , 5)$
10.	$x - y = 4$	$(0, \ ), (\ , 0), (6, \ ), (\ , -3)$
11.	$2x + y = 6$	$(0, \ ), (\ , 0), (4, \ ), (\ , 5)$
12.	$3x + 2y = 12$	$(0, \ ), (\ , 0), (6, \ ), \left(\ , -\dfrac{3}{2}\right)$
13.	$x - 4y = 8$	$(0, \ ), (\ , 0), (-4, \ ), (\ , 3)$
14.	$2x - 5y = 10$	$(0, \ ), (\ , 0), (-5, \ ), (\ , -3)$
15.	$7x + 3y = 21$	$(0, \ ), (\ , 0), (6, \ ), (\ , 14)$
16.	$8x + 5y = 40$	$(0, \ ), (\ , 0), (10, \ ), (\ , -4)$
17.	$x = -4$	$(\ , 0), (\ , -2), (\ , -5), (\ , 7)$
18.	$y = 2$	$(0, \ ), (4, \ ), (-1, \ ), (-9, \ )$
19.	$y + 4 = 0$	$(0, \ ), (-2, \ ), (8, \ ), (-4, \ )$
20.	$x - 5 = 0$	$(\ , 0), (\ , 9), (\ , 5), (\ , 2)$

Objective 3 Graph lines.

Objective 4 Find x- and y-intercepts.

Find the x-intercept and y-intercept of each line. (Note that Exercises 21–28 are graphed in Exercises 39–46.)

21. $x + y = 6$ **22.** $x + y = -3$ **23.** $x - y = 1$

24. $x - y = -7$

25. $3x + 2y = 6$

26. $5x + 2y = 10$

27. $5x - 4y = 20$

28. $7x - 3y = 21$

29. $11x + 4y = 22$

30. $5x + 9y = 18$

31. $3x + 7y = -8$

32. $4x - 7y = -5$

33. $x = 2$

34. $y = 5$

35. $y + 3 = 0$

36. $x + 4 = 0$

37. $3x - 4y = 0$

38. $3x + y = 0$

Graph each line.

39. $x + y = 6$

40. $x + y = -3$

41. $x - y = 1$

42. $x - y = -7$

43. $3x + 2y = 6$

44. $5x + 2y = 10$

45. $5x - 4y = 20$

46. $7x - 3y = 21$

47. $4x + y = -6$

48. $3x - 2y = -12$

49. $4x - 3y = 0$

50. $3x + 2y = 0$

Objective 5 **Recognize equations of vertical or horizontal lines.**

Graph each line.

51. $x = 2$

52. $y = 0$

53. $y = -3$

54. $x = -5$

55. $y - 4 = 0$

56. $x + 3 = 0$

57. $y + 6 = 0$

58. $x - 3 = 0$

4.1 Mixed Exercises

Complete the ordered pairs for the given equation.

Equation   *Ordered Pairs*

59. $3x - y = 3$ (, 0), (0,), (2,), (, 9)

60. $2x + 5y = 15$ (, 0), (0,), (-5,), (, 1)

61. $x + 4y = 8$ (, 0), (0,), (-1,), (, -3)

62. $2x - 3y = 0$ (, 0), (0,), (4,), (, 1)

For each equation, find the x-intercept, the y-intercept, and graph.

63. $x - y = 4$

64. $3x + y = 6$

65. $x = -4$

66. $3x - y = 0$

67. $y + 4 = 0$

68. $3x - 4y = 9$

4.2   Slope

Objective 1   Find the slope of a line given two points on the line.

Find the slope of the line through each pair of points using the slope formula.

1. $(5, 7), (7, 9)$

2. $(8, 2)(11, 5)$

3. $(9, -4), (7, -7)$

4. $(4, -2), (5, -3)$

5. $(-5, 7), (3, 23)$

6. $(3, -2), (-4, -7)$

7. $(1, -7), (3, -2)$

8. $(-4, 4), (2, -14)$

9. $(-3, -6), (-1, -7)$

10. $(-9, -3), (-5, -1)$

11. $(3, 6), (-2, 6)$

12. $(1, 9), (-2, 9)$

13. $(-6, -3), (0, 0)$

14. $(5, -4), (0, 0)$

Objective 2   Find the slope of a line given an equation of the line.

Find the slope of each line.

15. $y = 4x - 3$

16. $y = 2x + 7$

17. $y = 2 - x$

18. $y = 9 - 3x$

19. $x + y = 4$

20. $y = -6$

21. $4x + 3y = 12$

22. $3x - 2y = 12$

23. $9x + 5y = 18$

24. $7x - 3y = 4$

25. $2x - 7y = 1$

26. $x + 6y = 8$

27. $y - 4 = 0$

28. $x + 5 = 0$

29. $y = -2$

30. $x = 3$

Objective 3   Graph a line given its slope and a point on the line.

Graph each line.

	Slope	Through
31.	$\dfrac{2}{3}$	$(-1, 4)$
32.	$\dfrac{3}{4}$	$(2, 1)$
33.	$-\dfrac{1}{2}$	$(2, -3)$
34.	$-\dfrac{2}{3}$	$(-1, -2)$

	Slope	Through
35.	2	$(2, -4)$
36.	-1	$(-1, 2)$
37.	0	$(-2, 4)$
38.	undefined	$(-1, 0)$

Objective 4 **Use slopes to determine whether two lines are parallel, perpendicular, or neither.**

Decide whether each pair of lines is **parallel, perpendicular,** *or* **neither.**

39. $x + y = 6, x + y = -2$

40. $x - y = 1, x + y = 0$

41. $2x - y = 4, x + 2y = 9$

42. $3x + 2y = 4, 3x + 2y = 1$

43. $5x - y = 8, y = 5x + 1$

44. $6x - 5y = 9, 5x + 5y = 1$

45. $8x + y = 2, 8x - y = 1$

46. $4x - 3y = 2, 3x - 4y = 2$

47. the line through $(9, -2)$ and $(-1, 3)$, and the line through $(5, 7)$ and $(6, 5)$

48. the line through $(-7, 3)$ and $(2, -4)$, and the line through $(-8, 4)$ and $(-1, 13)$

Objective 5 **Solve problems involving average rate of change.**

Solve each problem.

49. Suppose the sales of a company are given by the linear equation $y = 1250x + 10{,}000$, where x is the number of years after 1980, and y is the sales in dollars. What is the average rate of change in sales per year?

50. Suppose a man's salary was \$15,750 in 1979 and has risen an average of \$1500 per year. If the yearly salaries were plotted on a graph, what would be the slope of the line on which they approximately lie?

51. A small company had the following sales during their first three years of operation.

Year	Sales
2003	\$82,250
2004	\$89,790
2005	\$96,100

(a) What was the rate of change from 2003–2004?

(b) What was the rate of change from 2004–2005?

(c) What was the average rate of change from 2003–2005?

52. A plane had an altitude of 8500 feet at 4:02 P.M. and 12,700 feet at 4:39 P.M. What was the average rate of change in the altitude in feet per minute?

53. A ramp is 10 feet high on the high end and 3 feet high on the low end. It covers a horizontal length of 29 feet. What is the average rate of change of the incline?

54. Enrollment in a college was 11,500 two years ago, 10,975 last year, and 10,800 this year.

(a) What is the average rate of change in enrollment per year for this 3–year period?

(b) Explain why the rate of change is negative.

4.2   Mixed Exercises

Find the slope of each line.

55. through (6, 3) and (−3, 6)

56. through (6, −2) and (−5, −2)

57. $2x + 3y = 6$

58. $5x − 2y = 10$

59. through (5, 7) and (5, −2)

60. $y = −2$

61. $y + x = 5$

62. through (3, 0), (0, −1)

Graph each line using the given slope and a point of the line.

63. slope 3, through (1, 2)

64. slope −1, through (3, 4)

65. slope 0, through (4, −2)

66. undefined slope, through (2, −3)

Decide whether each pair of lines is **parallel, perpendicular,** *or* **neither.**

67. $x = 6, x = −2$

68. $y = −3, y = 5$

69. The line through (−1, 7) and (2, 4), and the line through (7, 9) and (8, 8)

70. The line through (7, 6) and (9, −4), and the line through (8, 3) and (3, 2)

Solve each problem.

71. A company had 41 employees during the first year of operation. During their eighth year, the company had 79 employees. What was the average rate of change in the number of employees per year?

72. A state had a population of 3,105,900 in 1985. The population is declining at an average rate of 5200 people a year. At that rate, predict the population for the year 1999.

4.3 Linear Equations in Two Variables

Objective 1 Write the equation of a line given its slope and y-intercept.

Write an equation in standard form for each line.

	Slope	y-intercept			Slope	y-intercept
1.	2	$(0, -5)$		**7.**	$\frac{3}{5}$	$\left(0, \frac{2}{5}\right)$
2.	6	$(0, -2)$				
3.	-4	$(0, 3)$		**8.**	$\frac{6}{5}$	$\left(0, -\frac{1}{5}\right)$
4.	-5	$(0, 3)$				
5.	$-\frac{2}{3}$	$(0, 2)$		**9.**	$\frac{7}{3}$	$(0, 9)$
				10.	0	$(0, 3)$
6.	$-\frac{1}{4}$	$(0, -3)$				

Objective 2 Graph a line using its slope and *y*-intercept.

Objective 3 Write the equation of a line given its slope and a point on the line.

Write an equation for each line. Write answers in the form $Ax + By = C$.

	Slope	Point			Slope	Point
11.	2	$(-4, 1)$		**19.**	$-\frac{3}{4}$	$(-1, -3)$
12.	4	$(2, 6)$				
13.	5	$(3, -6)$		**20.**	$-\frac{4}{5}$	$(3, -2)$
14.	1	$(-4, 5)$				
15.	-5	$(-4, 19)$		**21.**	undefined	$(3, 0)$
16.	-3	$(2, -3)$		**22.**	0	$(-5, 2)$
				23.	0	$(3, -4)$
17.	$-\frac{1}{2}$	$(-3, 2)$		**24.**	undefined	$(0, 6)$
				25.	undefined	$(2, 7)$
18.	$-\frac{2}{3}$	$(1, -5)$		**26.**	0	$(3, 0)$

Objective 4 Write the equation of a line given two points on the line.

Write an equation in standard form of the line passing through the given pair of points.

27. $(4, 9), (3, 8)$

28. $(7, 1), (6, 5)$

29. $(3, 7), (5, 4)$

30. $(2, -1), (5, -2)$

31. $(-6, 2), (-4, 1)$

32. $(3, -2), (-1, 5)$

33. $(-3, -2), (-5, -1)$

34. $(-1, -4), (-2, -3)$

35. $(9, 1), (-9, 1)$

36. $(3, -5), (-4, -5)$

37. $(0, 2), (0, -6)$

38. $(-1, -7), (-1, 8)$

Objective 5 Write the equation of a line parallel or perpendicular to a given line.

Write an equation in standard form for each line.

39. parallel to $x - y = 4$, through $(4, -7)$

40. parallel to $2x + 3y = -12$, through $(9, -3)$

41. parallel to $2x + 6y = 5$, through $(1, -2)$

42. parallel to $4x - 3y = 8$, through $(-2, 3)$

43. parallel to $5x + y = 6$, through $(0, 4)$

44. perpendicular to $x - 3y = 0$, through $(-10, 2)$

45. perpendicular to $5x + y = 8$, through $(2, -1)$

46. perpendicular to $3x - 2y = 6$, through $(5, -3)$

47. perpendicular to $x = 4$, through $(-1, 7)$

48. perpendicular to $y = -1$, through $(2, 5)$

49. parallel to $y = 2$, through $(-4, 6)$

50. parallel to $x + 1 = 3$, through $(-3, 5)$

Objective 6 Write an equation of a line that models real data.

For each situation,
  (a) *Write an equation in the form $y = ax$;*
  (b) *Give the three ordered pairs associated with the equation for x-values 0, 5, and 10.*

51. x represents the number of minutes for a long distance call at $.13 per minute, and y represents the total cost of the call (in dollars).

52. x represents the number of rows of chairs, with 8 chairs in each row, set up for a concert, and y represents the total available seating.

53. x represents the number of lines of type in an ad at $1.25 per line, and y represents the total charge (in dollars).

For each situation,
  (a) *Write an equation in the form $y = ax + b$;*
  (b) *Give the three ordered pairs associated with the equation for x-values 0, 5, and 10.*

54. A long distance phone call costs $.35 plus $.13 per minute for each minute of the call. Let x represent the number of minutes so that y represents the total cost of the call (in dollars).

55. A music teacher set up rows of chairs for a concert. There were 8 chairs in each row, plus 15 special reserved seats up front for faculty. Let x represent the number of rows of chairs so that y represents the total number of guests who can be seated.

56. To run a newspaper ad, there is a $25 set up fee plus a charge of $1.25 per line of type in the ad. Let x represent the number of lines of type in the ad so that y represents the total cost of the ad (in dollars).

Write a linear equation and solve it in order to solve the problem.

57. Refer to Exercise 54. Suppose the call cost $1.91. How long was the call (in minutes)?

58. Refer to Exercise 55. Suppose 207 people were seated for the concert. How many rows of seats were there?

59. Refer to Exercise 56. A newspaper ad cost $62.50. How many lines long was the ad?

4.3   Mixed Exercises

Write an equation in standard form of each line satisfying the given conditions.

60. slope, $-\dfrac{5}{8}$; y-intercept, $\left(0, -\dfrac{2}{3}\right)$

61. through (2, 5) and (3, 6)

62. slope, $-\dfrac{4}{7}$; through (0, –8)

63. undefined slope; through (–3, 5)

64. parallel to $3x + 4y = 4$; through (–8, 4)

65. slope, 0; y-intercept, (0, –5)

66. slope, 0; through (0, 2)

67. through (–2, –4) and (–2, –7)

68. perpendicular to $x + y = 4$; through (2, 5)

69. through (9, –2) and (10, –5)

70. slope, –3; through (–4, 11)

71. through (6, 5) and (–6, 5)

Find the slope and y-intercept of each line.

72. $2x + 7y = 14$ **73.** $3x - 2y = 9$ **74.** $x = -10$ **75.** $y = 2$

4.4   Linear Inequalities in Two Variables

Objective 1   **Graph linear inequalities in two variables.**

Graph each linear inequality.

1. $x + y \geq 2$

4. $x - y < -4$

7. $3x - y \geq -3$

10. $y \geq -3$

2. $x + y \leq 5$

5. $2x + 3y \geq 6$

8. $2x - 5y < -10$

3. $x - y < 5$

6. $3x - 2y \leq 12$

9. $x \leq 4y$

Objective 2   **Graph the intersection of two linear inequalities**

Graph the intersection of each pair of inequalities.

11. $x + y \leq 4$ and $x - y \geq 2$

15. $3x + 4y \leq 12$ and $2x - y \leq 4$

12. $x - y < 3$ and $x + y > -2$

16. $5x + 2y < 10$ and $2x + 3y > 6$

13. $2x + y < 6$ and $x - 3y > -6$

17. $2x + y \geq 6$ and $y \geq 2$

14. $4x + y \leq 4$ and $x - 2y \leq -2$

18. $4y - 3x \leq 12$ and $x \geq 0$

Objective 3   **Graph the union of two linear inequalities.**

Graph the union of each pair of inequalities.

19. $4x - 2y \geq -4$ or $x \geq 1$

23. $x \geq 4$ or $y < -3$

20. $x - 4y \leq -2$ or $x \leq 3$

24. $y \geq 2x$ or $x \geq 0$

21. $4x - 2y \geq 8$ or $y \geq 2$

25. $x + y \geq 0$ or $x - y \geq 0$

22. $2x + y < -1$ or $x - 2y > 1$

26. $y > 4$ or $y < -4$

4.4   Mixed Exercises

Graph the solution set of each of the following.

27. $y + 4x < 0$

31. $x > 2$ and $2x - 3y < 6$

28. $x - y \leq 4$ and $x + y \geq -3$

32. $x - 2y \geq 0$

29. $x + 2y > 2$ or $y \leq 0$

33. $x \leq -2$

30. $y > 3x$

34. $x + y \geq 2$ or $2x - y \leq 4$

4.5   Introduction to Functions

Objective 1   Define relation and function.

Decide whether each relation is a function.

1. $\{(1, 3), (1, 4), (2, -1), (3, 7)\}$

2. $\{(-1, 2), (0, 5), (1, 8)\}$

3. $\{(2, -2), (3, -3), (4, -4)\}$

4. $\{(6, -3), (4, -2), (2, -1), (0, 0)\}$

5. $\{(0, 4), (3, 2), (0, 0), (3, 5)\}$

6. $\{(-4, -1), (-3, -2), (-1, 0), (0, 5)\}$

7. $\{(1, 1), (1, 2), (1, 7), (2, 1)\}$

8. $\{(3, 4), (5, 2), (4, 3), (5, 3)\}$

9. $\{(1, 5), (2, 5), (4, 5)\}$

10. $y + x = 9$

11. $x^2 + y^2 = 1$

12. $y \geq 8x - 3$

13. $xy = 7$

14. $x = \sqrt{y + 1}$

15. $x = y$

Objective 2   Find domain and range.

Decide whether each relation is a function and give the domain and range of the relation.

16. $\{(-1, 1), (-2, 2), (0, 0)\}$

17. $\{(3, 0), (2, 4), (1, 6), (-1, 3)\}$

18. $\{(-2, -2), (-1, -1), (0, 0), (1, -1)\}$

19. $\{(3, 5), (2, 3), (1, 0)\}$

20. $\{(1, 3), (2, -1), (-1, 4), (1, 4)\}$

21. $\{(2, -4), (1, -2), (-1, 2), (0, 3)\}$

22. $\{(5, 2), (3, -1), (1, -3), (-1, -5)\}$

23. $\{(4, 2), (3, 2), (2, 2), (1, 2), (0, 2)\}$

Decide whether each relation defines y as a function of x. Give the domain.

24. $y = 2x + 5$

25. $y \leq 4x$

26. $y = \sqrt{x - 4}$

27. $y^2 = x + 1$

28. $y = \dfrac{1}{x}$

29. $y = 2x^2 + 3$

Objective 3   Identify functions.

Decide which are graphs of functions. Use the vertical line test.

30.

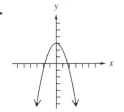

31.

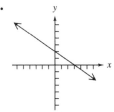

32.

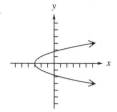

33.

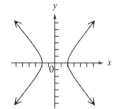

34.

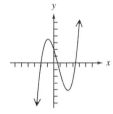

35.

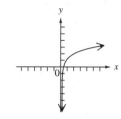

36.

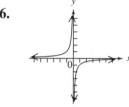

37.

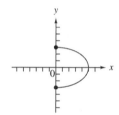

Objective 4   **Use function notation.**

Find $f(-2)$, $f(4)$, and $f(-x)$.

38. $f(x) = 2x + 5$

39. $f(x) = 6 - 2x$

40. $f(x) = 3x^2$

41. $f(x) = x^2 - 2x$

42. $f(x) = \dfrac{4}{x^2 + 1}$

43. $f(x) = \dfrac{2x + 1}{5}$

Objective 5   **Identify linear functions.**

Write each equation in the form $f(x) = mx + b$.

44. $y = x + 2$

45. $2x - y = -2$

46. $x - y = 1$

47. $y + \dfrac{1}{2}x = -2$

48. $x + 4y = 2$

49. $\dfrac{1}{3}y = x$

50. $x + y = 1$

51. $\dfrac{1}{2}x + \dfrac{1}{3}y = -1$

4.5   Mixed Exercises

Decide whether each relation is a function, and give the domain and range of the relation.

52. $\{(0, 1), (1, 3), (2, -4), (4, -8)\}$

53. $\{(-2, -5), (-3, -2), (1, -2)\}$

54. $\{(-4, 1), (-5, 2), (-4, 3), (-3, 4), (-2, 5)\}$

55. $\{(1, 10), (2, 9), (3, 8), (4, 7), (5, -4)\}$

Decide whether the given equation defines y as a function of x. Give the domain.

56. $y = x - 6$

58. $y = |x|$

57. $y = \sqrt{x + 2}$

59. $y^2 - 4 = x$

Write each equation in the form $f(x) = mx + b$. Then find $f(-1), f(6),$ and $f(x - 1)$.

60. $3y + x = 9$

61. $x = \dfrac{1}{2}y + 6$

4.6   Variation

Objective 1   **Write an equation expressing direct variation.**

Objective 2   **Find the constant of variation, and solve direct variation problems.**

Suppose y varies directly as x. Find the equation connecting y and x.

1. $y = 15$ when $x = 5$.
      6. $y = 14$ when $x = 9$.

2. $y = 30$ when $x = 6$.
      7. $y = 150$ when $x = 3$.

3. $y = 12$ when $x = 8$.
      8. $y = 270$ when $x = 9$.

4. $y = 11$ when $x = 22$.
      9. $y = 8$ when $x = 6$.

5. $y = 23$ when $x = 12$.
      10. $y = 13.75$ when $x = 55$.

Solve each problem.

11. If r varies directly as t, and $r = 10$ when $t = 2$, find r when $t = 9$.

12. If q varies directly as p, and $q = 36$ when $p = 5$, find q when $p = 20$.

13. If x varies directly as y, and $x = 9$ when $y = 2$, find x when y is 7.

14. If z varies directly as x, and $z = 15$ when $x = 4$, find z when x is 11.

15. If m varies directly as p^2, and $m = 20$ when $p = 2$, find m when p is 5.

16. If a varies directly as b^2, and $a = 48$ when $b = 4$, find a when $b = 7$.

17. If y varies directly as the square of z, and $y = 8$ when $z = 6$, find y when $z = 9$.

18. If m varies directly as the square of p, and $m = 1$ when $p = 2$, find m when p is 7.

19. The circumference of a circle varies directly as the radius. A circle with a radius of 7 centimeters has a circumference of 43.96 centimeters. Find the circumference of the circle if the radius changes to 11 centimeters.

20. The pressure exerted by a certain liquid at a given point varies directly as the depth of the point beneath the surface of the liquid. The pressure at 10 feet is 50 pounds per square inch (psi). What is the pressure at 20 feet?

21. The force required to compress a spring varies directly as the change in length of the spring. If a force of 20 newtons is required to compress a spring 2 centimeters in length, how much force is required to compress a spring of length 10 centimeters?

22. The surface area of a sphere varies directly as the square of its radius. If the surface area of a sphere with a radius of 12 inches is 576π square inches, find the surface area of a sphere with a radius of 3 inches.

Objective 3   **Solve inverse variation problems.**

Suppose y varies inversely as x. Find the equation connecting y and x.

23. $y = 10$ when $x = 2$.

24. $y = 4$ when $x = 6$.

25. $y = 8$ when $x = 10$.

26. $y = 2$ when $x = 12$.

27. $y = 5$ when $x = \dfrac{1}{6}$.

28. $y = 9$ when $x = \dfrac{3}{2}$.

Solve each problem.

29. If y varies inversely as x, and $y = 10$ when $x = 3$, find y when $x = 12$.

30. If z varies inversely as y^2, and $z = 12$ when $y = 3$, find z when $y = 6$.

31. If p varies inversely as q^2, and $p = 4$ when $q = \frac{1}{2}$, find p when $q = \frac{3}{2}$.

32. If z varies inversely as x^2, and $z = 9$ when $x = \frac{2}{3}$, find z when $x = \frac{5}{4}$.

33. The current in a simple electrical circuit varies inversely as the resistance. If the current is 50 amperes (an ampere is a unit for measuring current) when the resistance is 10 ohms (an ohm is a unit for measuring resistance), find the current if the resistance is 5 ohms.

34. The illumination produced by a light source varies inversely as the square of the distance from the source. If the illumination produced 4 feet from a light source is 75 footcandles, find the illumination produced 9 feet from the same source.

35. The weight of an object varies inversely as the square of its distance from the center of the earth. If an object 8000 miles from the center of the earth weighs 90 pounds, find its weight when it is 12,000 miles from the center of the earth.

36. The speed of a pulley varies inversely as its diameter. One kind of pulley, with a diameter of 3 inches, turns at 150 revolutions per minute. Find the speed of a similar pulley with diameter of 5 inches.

Objective 4   **Solve joint variation problems.**

Suppose y varies jointly as x and z. Find the equation connecting y, x, and z.

37. $y = 10$ when $x = 2$ and $z = 5$.

38. $y = 50$ when $x = 4$ and $z = 6$.

39. $y = 27$ when $x = 9$ and $z = 6$.

40. $y = 144$ when $x = 8$ and $z = 9$.

41. $y = 24$ when $x = 12$ and $z = 4$.

42. $y = 30$ when $x = 10$ and $z = 12$.

Solve each problem.

43. If r varies jointly as m and n^2, and $r = 72$ when $m = 4$ and $n = 6$, find r when $m = 3$ and $n = 4$.

44. If q varies jointly as p and r^2, and $q = 27$ when $p = 9$ and $r = 2$, find q when $p = 8$ and $r = 4$.

45. Suppose y varies jointly as x^2 and z^2, and $y = 72$ when $x = 2$ and $z = 3$. Find y when $x = 4$ and $z = 2$.

46. Suppose d varies jointly as f^2 and g^2, and $d = 384$ when $f = 3$ and $g = 8$. Find d when $f = 6$ and $g = 2$.

47. For a fixed period of time, interest varies jointly as the principal and the rate of interest. If a principal of $2000 invested at a rate of 6% earns $300 in interest, how much interest will a principal of $4000 earn if it is invested at 8% for the same period of time?

48. The absolute temperature of an ideal gas varies jointly as its pressure and its volume. If the absolute temperature is 250° when the pressure is 25 pounds per square centimeter and the volume is 50 cubic centimeters, find the absolute temperature when the pressure is 50 pounds per square centimeter and the volume is 75 cubic centimeters.

Objective 5   **Solve combined variation problems.**

Suppose y varies directly as x and inversely as z. Find the equation connecting y, x, and z.

49. $y = 1$ when $x = 2$ and $z = 6$.

50. $y = 6$ when $x = 4$ and $z = 8$.

51. $y = 2$ when $x = 6$ and $z = 2$.

52. $y = 3$ when $x = 72$ and $z = 8$.

53. $y = .1625$ when $x = 4$ and $z = 16$.

54. $y = 10.5$ when $x = 1.8$ and $z = .6$.

Solve each problem.

55. The time required to lay a sidewalk varies directly as its length and inversely as the number of people who are working on the job. If three people can lay a sidewalk 100 feet long in 15 hours, how long would it take two people to lay a sidewalk 40 feet long?

56. When an object is moving in a circular path, the centripetal force varies directly as the square of the velocity and inversely as the radius of the circle. A stone that is whirled at the end of a string 50 centimeters long at 900 centimeters per second has a centripetal force of 3,240,000 dynes. Find the centripetal force if the stone is whirled at the end of a string 75 centimeters long at 1500 centimeters per second.

4.6   Mixed Exercises

Solve each problem.

57. If p varies directly as q, and $p = 14$ when $q = 2$, write an equation connecting p and q.

58. If w varies jointly as v and s, and $w = 3.5$ when $v = 25$ and $s = 7$, write an equation connecting w, v, and s.

59. If r varies inversely as s, and $r = 7$ when $s = 8$, find r when $s = 12$.

60. If m varies directly as w and inversely as r^2, and $m = 1845$ when $w = 4.5$ and $r = .1$, find m when $w = 2.5$ and $r = .2$.

61. If r varies inversely as t^2, and $r = 8$ when $t = 4$, find r when $t = 9$.

62. p varies jointly as P, V, and t and inversely as v and T. Suppose $p = 65.625$ when $P = 50$, $V = 9$, $t = 350$, $v = 8$, and $T = 300$. Find p when $P = 60$, $V = 8$, $t = 300$, $v = 6$, and $T = 200$.

63. For a body falling free from rest (disregarding air resistance), the distance the body falls varies directly as the square of the time. If an object is dropped from the top of a tower 400 feet high and hits the ground in 5 seconds, how far did it fall in the first 3 seconds?

64. The resistance in ohms of a platinum wire temperature sensor varies directly as the temperature in degrees Kelvin (K). If the resistance is 646 ohms at a temperature of 190 K, find the resistance at a temperature of 250 K.

65. The volume of a gas varies inversely as the pressure and directly as the temperature. If a certain gas occupies a volume of 1.3 liters at 300 K and a pressure of 18 kilograms per square centimeter, find the volume at 340 K and a pressure of 24 kilograms per square centimeter.

66. The force required to compress a spring varies directly as the change in the length of the spring. If a force of 12 pounds is required to compress a certain spring 3 inches, how much force is required to compress the spring 5 inches?

67. The distance that a person can see to the horizon on a clear day from a point above the surface of the earth varies directly as the square root of the height at the point. If a person 144 meters above the surface can see 18 kilometers to the horizon, how far can a person see to the horizon from a point 64 meters above the surface?

68. For a fixed interest rate, interest varies jointly as the principal and the time in years. If a principal of $5000 invested for 4 years earns $900, how much interest will $6000 invested for 3 years earn at the same interest rate?

Chapter 5

SYSTEMS OF LINEAR EQUATIONS

5.1   Systems of Linear Equations in Two Variables

Objective 1   Solve linear systems by graphing.

Solve each system by graphing.

1. $x + y = 3$
   $x - y = -1$

2. $x + y = 5$
   $x - y = 3$

3. $x - y = -2$
   $3x + 2y = -6$

4. $x + 4y = 1$
   $2x + y = 2$

5. $x + 2y = 4$
   $3x + y = -3$

6. $x - 2y = 1$
   $x + 4y = 7$

7. $x + 2y = 0$
   $2x + y = -6$

8. $x + y = 2$
   $2x + 5y = 10$

Objective 2   Decide whether an ordered pair is a solution of a linear system.

Decide whether the ordered pair is a solution for the given system.
Write solution *or* not a solution.

9. $2x - 5y = 5$ $(5, 1)$
   $4x + 3y = 23$

10. $4x + 5y = 11$ $(-1, 3)$
    $3x - y = -6$

11. $4x - 5y = 26$ $(4, -2)$
    $3x + 2y = 6$

12. $5x - 3y = -19$ $(-2, 3)$
    $2x + 3y = 6$

13. $7x + 9y = 15$ $(6, -3)$
    $4x + 8y = 21$

14. $3x - y = 7$ $(3, 2)$
    $4x - 5y = 2$

15. $x = 4y$ $(12, 3)$
    $3x = y + 33$

16. $5y = x$ $(15, 3)$
    $2y = x - 9$

17. $2x - 9y = 27$ $(0, -3)$
    $5x + 6y = 18$

18. $7x + 8y = -14$ $(-2, 0)$
    $5x - 7y = 10$

Objective 3   Solve linear systems by substitution.

Solve each system by the substitution method.

19. $3x + y = -20$
    $y = 2x$

20. $4x - y = -7$
    $y = -3x$

21. $x + 2y = 5$

    $x = 2y + 1$

22. $4x - 3y = 15$

    $x = y + 4$

23. $3x + 7y = 16$

    $y = 2x - 5$

24. $2x + y = 6$

    $y = 5 - 3x$

25. $y = 11 - 2x$

    $x = 18 - 3y$

26. $y = 2x + 5$

    $y = -4x + 2$

27. $x + 6y = 15$

    $y = \dfrac{2}{3}x$

28. $3x - 2y = -1$

    $x = \dfrac{3}{4}y$

29. $\dfrac{x}{2} + \dfrac{y}{3} = \dfrac{3}{2}$

    $y = 3x$

30. $\dfrac{x}{4} - \dfrac{y}{5} = \dfrac{3}{4}$

    $y = 5x$

31. $3x + 2y = 7$

    $4x - 3y = -19$

32. $4x + 5y = 13$

    $3x - 4y = 2$

33. $x + 3y = 9$

    $x - 2y = -1$

Objective 4   **Solve linear systems by elimination method.**

Use the elimination method to solve each system of linear equations.

34. $x + 2y = 7$

    $x - y = -2$

35. $3x - y = 11$

    $x + y = 5$

36. $5x + 2y = 16$

    $3x - 4y = 20$

37. $3x + 4y = -13$

    $5x - 2y = -13$

38. $8x + 3y = 13$

    $3x + 2y = 11$

39. $2x + 8y = 3$

    $4x - 12y = -1$

40. $3x + 4y = 3$

    $9x - 8y = 4$

41. $\dfrac{1}{2}x + \dfrac{1}{4}y = 5$

    $\dfrac{1}{2}x - \dfrac{3}{4}y = -3$

42. $\dfrac{x}{5} + \dfrac{y}{2} = 0$

    $\dfrac{3x}{5} - \dfrac{5y}{2} = -8$

43. $\dfrac{x}{2} + \dfrac{y}{3} = -2$

    $\dfrac{3x}{2} + \dfrac{5y}{3} = -8$

Objective 5   **Solve special systems.**

44. $2x + y = 3$
$4x + 2y = 12$

45. $3x - y = 8$
$-6x + 2y = 16$

46. $5x - 2y = 4$
$10x - 4y = 8$

47. $9x - y = 6$
$-18x + 2y = -12$

48. $4x - 3y = 6$
$8x - 6y = 10$

5.1   Mixed Exercises

Solve each system by any method.

49. $8x + 7y = 52$
$x - 3 = 0$

50. $6x + 5y = 13$
$y + 1 = 0$

51. $3x + 2y = -8$
$4 + y = 2x$

52. $3x - 4y = 13$
$2x + 3y = 3$

53. $\dfrac{x}{6} + \dfrac{y}{4} = 1$
$4x + 6y = 24$

54. $3x + 7y = 32$
$y = \dfrac{1}{3}x$

55. $2x + 3y = 1$
$3x - 2y = 8$

56. $4x - 3y = -12$
$x + 3 = y$

57. $6x - 5y = -8$
$x = \dfrac{1}{2}y$

58. $\dfrac{x}{5} - \dfrac{y}{3} = \dfrac{3}{5}$
$-3x + 5y = 9$

59. $5x + 15y = 5$
$x + 3y = 18$

60. $-x + 2y = -8$
$2x - 4y = 16$

5.2 Systems of Linear Equations in Three Variables

Objective 2 Solve linear systems by elimination.

Solve each system of equations.

1. $x + y + z = 0$
$x - y + z = -2$
$x - y - z = -4$

2. $x + 2y + z = 4$
$2x + y - z = -1$
$-x + y + z = 2$

3. $x + y + z = 4$
$x - y + 2z = 8$
$2x + y - z = 3$

4. $3x - y + 2z = -6$
$2x + y + 2z = -1$
$3x + y - z = -10$

5. $4x + 2y + 3z = 11$
$2x + y - 4z = -22$
$3x + 3y + z = -1$

6. $x - 2y + 5z = -7$
$2x + 3y - 4z = 14$
$3x - 5y + z = 7$

7. $2x - 5y + 2z = 30$
$x + 4y + 5z = -7$
$\frac{1}{2}x - \frac{1}{4}y + z = 4$

8. $5x - 2y + z = 28$
$3x + 5y - 2z = -23$
$\frac{2}{3}x + \frac{1}{3}y + z = 1$

9. $\frac{1}{3}x + \frac{1}{6}y - \frac{2}{3}z = -1$
$\frac{3}{4}x + \frac{1}{3}y + \frac{1}{4}z = -3$
$\frac{1}{2}x + \frac{3}{2}y + \frac{3}{4}z = 21$

10. $\frac{2}{3}x - \frac{1}{4}y + \frac{5}{8}z = 0$
$\frac{1}{5}x + \frac{2}{3}y - \frac{1}{4}z = -7$
$\frac{3}{5}x - \frac{4}{3}y + \frac{7}{8}z = 5$

Objective 3 Solve linear systems where some of the equations have missing terms.

Solve each system of equations.

11. $x - z = -3$
$y + z = 4$
$x - y = 3$

12. $2x + 3y = 3$
$6y - 5z = 3$
$4x + 9y = 8$

13. $x + 5y = -23$
$4y - 3z = -29$
$2x + 5z = 19$

14. $3x - 4z = -23$
$y + 5z = 24$
$x - 3y = 2$

15. $4x - 5y = -13$
$3x + z = 9$
$2y + 5z = 10$

16. $7x + z = 5$
$3y - 2z = -16$
$5x + y = -2$

17. $2x + 5y \quad = 18$

$\quad\quad\; 3y + 2z = 4$

$\quad \dfrac{1}{4}x - y \quad = -1$

18. $5x - \quad\quad 2z = \; 8$

$\quad\quad\; 4y + 3z = -9$

$\quad \dfrac{1}{2}x + \dfrac{2}{3}y \quad = -1$

19. $\quad x + 2y \quad = -2$

$\quad\quad \dfrac{1}{2}y + z = -1$

$\quad \dfrac{2}{3}x - \dfrac{3}{4}y \quad = \; 7$

20. $\quad 4x - \quad\quad z = -6$

$\quad\quad \dfrac{3}{5}y + \dfrac{1}{2}z = \; 0$

$\quad \dfrac{1}{3}x + \quad\quad \dfrac{2}{3}z = -5$

Objective 4   **Solve special systems.**

Solve each systems of equations.

21. $\quad x - y + z = 7$

$\quad 2x + 5y - 4z = 2$

$\quad -x + y - z = 4$

22. $\quad 8x - 7y + 2z = 1$

$\quad 3x + 4y - z = 6$

$\quad -8x + 7y - 2z = 5$

23. $\quad 3x - 2y + 4z = \; 5$

$\quad -3x + 2y - 4z = -5$

$\quad \dfrac{3}{2}x - y + 2z = \; \dfrac{5}{2}$

24. $\quad -x + 5y - 2z = \; 3$

$\quad 2x - 10y + 4z = -6$

$\quad -3x + 15y - 6z = \; 9$

25. $\quad 8x - 4y + 2z = 0$

$\quad 3x + y - 4z = 0$

$\quad 5x + y + 2z = 0$

26. $\quad x - 3y + 4z = 0$

$\quad 2x + y - z = 0$

$\quad -x + y - 5z = 0$

27. $\quad 3x - 2y + 5z = \; 6$

$\quad x - 4y - z = \; 1$

$\quad \dfrac{3}{2}x - y + \dfrac{5}{2}z = -3$

28. $\quad 2x + 7y - 8z = 3$

$\quad 5x - y - z = 1$

$\quad x + \dfrac{7}{2}y - 4z = 3$

29. $\quad x - 5y + 2z = 0$

$\quad -x + 5y - 2z = 0$

$\quad \dfrac{1}{2}x - \dfrac{5}{2}y + z = 0$

30. $\quad 3x - 2y + 5z = 0$

$\quad 6x - 4y + 10z = 0$

$\quad \dfrac{3}{2}x - y + \dfrac{5}{2}z = 0$

5.2   Mixed Exercises

Solve each system of equations.

31. $x + y + z = 6$
$2x - y + z = 3$
$x + 2y - z = 2$

32. $3x - y + 2z = 9$
$2x + y - z = 7$
$x + 2y - 3z = 4$

33. $2x + y - 4z = 17$
$3x - 2y - z = -7$
$-2x + 2y + z = 7$

34. $6x + 2y - 4z = 6$
$-12x - 4y + 8z = -12$
$-3x - y + 2z = -3$

35. $2x - 3y = -2$
$x + y - 4z = -16$
$3x - 2y + z = 7$

36. $9x - 6z = 8$
$3x + 4y - 2z = 16$
$-6x + 4z = 11$

37. $2x + 5y + 3z = -1$
$3x - y + 2z = 7$
$4x + 2y + 3z = 6$

38. $5x - 2y + 5z = 27$
$4x + 3y + 4z = 17$
$2x - 4y - 3z = -1$

39. $3x - y = -2$
$y + 5z = -4$
$-2x + 3y - z = -8$

40. $3x - 4y + 2z = 3$
$x + 3y + 5z = 14$
$2x - 5y + 4z = 5$

5.3   Applications of Systems of Linear Equations

Objective 1   **Solve problems using two variables.**

For each word problem, select variables to represent the two unknowns, write two equations using the two variables, and solve the resulting system.

1. The length of a rectangle is 5 feet more than the width. The perimeter of the rectangle is 58 feet. Find the width of the rectangle.

2. The length of a rectangle is 7 centimeters more than the width. The perimeter of the rectangle is 134 centimeters. Find the width of the rectangle.

3. The perimeter of a rectangle is 70 inches. If the width were doubled, the width would be 10 inches more than the length. Find the width of the rectangle.

4. The perimeter of a rectangle is 96 inches. If the width were tripled, the width would be 36 inches more than the length. Find the width of the rectangle.

5. The width of a rectangle is 42 yards less than the length. The perimeter of the rectangle is 220 yards. Find the length of the rectangle.

6. The measure of the largest angle of a triangle is 45° more than the measure of the smallest angle. The middle angle measures 65°. What is the largest angle?

7. The side of a square is 4 centimeters longer than the side of an equilateral triangle. The perimeter of the square is 28 centimeters longer than the perimeter of the triangle. Find the length of a side of the triangle.

8. The perimeter of a triangle is 70 centimeters. Two sides of the triangle have the same length. The third side is 7 centimeters longer than either of the equal sides. Find the length of the equal sides of the triangle.

Objective 2   **Solve money problems using two variables.**

For each word problem, select variables to represent the two unknowns, write two equations using the two variables, and solve the resulting system.

9. Irene Doo has some $5-bills and some $10-bills. The total value of the money is $300, with a total of 40 bills. How many of each are there?

10. Pablo Gomez has some $10-bills and some $20-bills. The total value of the money is $650, with a total of 40 bills. How many of each are there?

11. Big Giant Super Market will sell 5 large jars and 2 small jars of their peanut butter for $19. They will also sell 2 large jars and 5 small jars for $16. What is the price of each jar?

12. The cost of 12 small oranges and 30 large oranges is $18.60. The cost of 24 small oranges and 12 large oranges is $13.20. Find the cost of a small orange and the cost of a large one.

13. The Garden Center ordered 6 ounces of marigold seed and 8 ounces of carnation seed, paying $214.54. They later ordered another 12 ounces of marigold seed and 18 ounces of carnation seed, paying $464.28. Find the price per ounce for each type of seed.

14. Jay bought 105 pounds of cattle feed and 62 pounds of rabbit feet, paying $109.75. He later bought 70 pounds of cattle feed and 35 pounds of rabbit feed for $70. Find the cost per pound for each type of feed.

15. A total of $40,000 is invested, part at 6% and part at 4%. The annual return is $1740. How much is invested at each rate?

16. Rachel has $20,000 to invest. She plans to invest part at 5%, with the remainder invested at 6%. Find the amount invested at each rate if the total annual interest income is $1060.

17. A taxi charges a flat rate plus a certain charge per mile. A trip of 4 miles costs $2.05, while a trip of 8 miles costs $2.85. Find the flat rate and the charge per mile.

18. A taxi charges a flat rate plus a certain charge per mile. A trip of 7 miles costs $2.65, while a trip of 3 miles costs $2.25. Find the flat rate and the charge per mile.

Objective 3   Solve mixture problems using two variables.

For each word problem, select variables to represent the two unknowns, write two equations using the two variables, and solve the resulting system.

19. How many ounces each of 20% acid and 50% acid must be mixed together to get 120 ounces of a 30% acid?

20. How many ounces each of 40% acid and 80% acid must be mixed together to get 160 ounces of a 70% acid?

21. A radiator holds 10 liters. How much pure antifreeze must be added to a mixture that is 10% antifreeze to make enough of a 20% mixture to fill the radiator?

22. A radiator holds 15 liters. How much pure antifreeze must be added to a mixture that is 10% antifreeze to make enough of a 25% mixture to fill the radiator?

23. How many pounds of an alloy containing 32% silver must be melted with 25 pounds of an alloy containing 48% silver to obtain an alloy containing 42% silver?

24. A candy mix is to be made by mixing candy worth $12 per kilogram with candy worth $15 per kilogram to get 120 kilograms of a mixture worth $13 per kilogram. How many kilograms of each should be used?

25. A coffee mix is to be made by mixing coffee worth $10 per kilogram with coffee worth $18 per kilogram to get 80 kilograms of a mixture worth $12 per kilogram. How many kilograms of each should be used?

26. Nuts worth $4 a pound are to be mixed with nuts worth $7 a pound to get 60 pounds of a mix worth $6 per pound. How many pounds of each will be needed?

Objective 4   Solve distance-rate-time problems using two variables.

For each word problem, select variables to represent the two unknowns, write two equations using the two variables, and solve the resulting system.

27. Two cars start together and travel in the same direction, one going twice as fast as the other. At the end of 3 hours, they are 96 miles apart. How fast is each traveling?

28. A plane flying with the wind flew 600 miles in 5 hours. Against the wind, the plane required 6 hours to fly the same distance. Find the rate of the wind.

29. Two trains start at the same point going in the same direction on parallel tracks. One train travels at 70 miles per hour and the other at 42 miles per hour. In how many hours will they be 154 miles apart?

30. Travis and his sister Kate jog to school daily. Travis jogs at 9 miles per hour, and Kate jogs at 5 miles per hour. When Travis reaches school, Kate is $\frac{1}{2}$ mile from the school. How far do Travis and Kate live from their school?

31. A motorboat traveling with the current went 72 miles in 3 hours. Against the current, the boat could only go 48 miles in the same amount of time. Find the rate of the boat in still water.

32. A plane can travel 600 miles per hour with the wind but in the same time only 530 miles per hour against the wind. Find the speed of the plane in still air.

33. Ashley walks 10 miles in the same time that Taylor walks 6 miles. If Ashley walks 1 mile per hour less than twice Taylor's rate, what is the rate at which Ashley walks?

34. An airplane flying with the wind travels from Palatine to Antioch, a distance of 500 miles, in 2 hours. The return trip, against the wind, takes $2\frac{1}{2}$ hours. What is the average speed of the plane in still air?

35. Determine the rate of the river current, if it takes Lynn 2 hours to row 9 miles with the current and 6 hours to return the same distance against the current.

36. A train travels 600 kilometers in the same time that a truck travels 520 kilometers. Find the speed of the train if the train's average speed is 8 kilometers per hour faster than the truck's.

| Objective 5 | **Solve problems with three variables using a system of three equations.**

For each problem, select variables to represent the three unknowns, write three equations using the three variables, and solve the resulting system.

37. Three numbers have a sum of 31. The middle number is 1 more than the smallest number. The sum of the smaller two numbers is 7 more than the largest number. Find the three numbers.

38. The sum of three numbers is 99. The difference of the smaller two is 3. The sum of the smallest and twice the largest is 108. Find the three numbers.

39. The sum of the measures of the angles of any triangle is 180°. In a certain triangle, the first angle measures 20° less than the second angle, and the second angle measures 10° more than the third. Find the measures of the three angles.

40. In a certain triangle, the sum of the measures of the smallest and largest angles is 50° more than the measure of the medium angle. The medium angle measures 25° more than the smallest angle. Find the measures of the three angles.

41. The perimeter of a triangle is 60 inches. The second side is 10 inches longer than the first side, and the third side is 6 inches more than twice the first side. Find the lengths of the sides.

42. Lance has some $5, $10, and $20-bills. He has a total of 51 bills, worth $795. The number of $5-bills is 25 less than the number of $20-bills. Find the number of each type of bill he has.

43. Emily has some $10, $20, and $50-bills. She has a total of 50 bills, worth $1270. The number of $20 bills equals the combined number of $10 and $50-bills. Find the number of each type of bill.

44. Sara Mitchell has $80,000 to invest. She invests part at 5%, one fourth this amount at 6%, and the balance at 7%. Her total annual income from interest is $4700. Find the amount invested at each rate.

45. A company borrowed a total of $75,000. Some of the money was borrowed at 8% interest, and $30,000 more than that amount was borrowed at 10%. The rest was borrowed at 11%. How much was borrowed at each rate if the total annual simple interest was $7150?

46. A merchant wishes to mix gourmet coffee selling for $8 per pound, $10 per pound, and $15 per pound to get 50 pounds of a mixture that can be sold for $11.70 per pound. The amount of the $8 coffee must be 3 pounds more than the amount of the $10 coffee. Find the number of pounds of each that must be used.

5.3   Mixed Exercises

Use a system of equations to solve each problem.

47. The side of a square is 5 centimeters shorter than the side of an equilateral triangle. The perimeter of the square is 7 centimeters less than the perimeter of the triangle. Find the lengths of a side of the square and of a side of the triangle.

48. Philip bought 2 kilograms of dark clay and 3 kilograms of light clay, paying $13 for the clay. He later needed 1 kilogram of dark clay and 2 kilograms of light clay, costing $7 altogether. How much did he pay for each type of clay?

49. A total of $12,000 is invested, part at 6% simple interest and part at 3%. If the annual return from the two investments is the same, how much is invested at each rate?

50. The cost of a ticket to a concert was $10 if purchased prior to the day of the concert. However, if purchased the day of the concert, the cost of the ticket increased to $12. The receipts were $39,600, and a total of 3300 tickets was sold. How many tickets were purchased the day of the concert?

51. The perimeter of a triangle is 81 centimeters. The smallest side is 7 centimeters shorter than the medium side. The medium side is 4 centimeters shorter than the longest side. Find the lengths of the three sides.

52. How many gallons each of 15% alcohol and 30% alcohol should be mixed to get 30 gallons of 25% alcohol?

53. A car travels 225 kilometers in the same time that a truck travels 195 kilometers. If the speed of the car is 10 kilometers per hour faster than the speed of the truck, find the speed of each vehicle.

54. The manager of the Sweet Candy Shop wishes to mix candy worth $4 per pound, $6 per pound, and $10 per pound to get 100 pounds of a mixture worth $7.60 per pound. The amount of $10 candy must equal the total amounts of the $4 and the $6 candy. How many pounds of each must be used?

55. In his motorboat, Luke travels downstream a distance of 21 miles in 1 hour. Returning, he finds that the trip upstream takes $1\frac{2}{5}$ hours. What is the speed of the current? What is the speed of Luke's boat in still water?

56. A box contains $6.25 in nickels, dimes, and quarters. There are 85 coins in all, with three times as many nickels as dimes. How many quarters are there?

5.4   Solving Systems of Linear Equations by Matrix Methods

Objective 1   **Define a matrix.**

Give the dimensions of each matrix.

1. $\begin{bmatrix} 1 & 2 \\ -1 & 0 \end{bmatrix}$

2. $\begin{bmatrix} -4 & 3 \\ 0 & 7 \\ 6 & 2 \end{bmatrix}$

3. $\begin{bmatrix} 1 & 4 & 7 \\ 6 & 5 & -5 \end{bmatrix}$

4. $\begin{bmatrix} 1 & 4 \\ 3 & 2 \\ -2 & 0 \\ 5 & -3 \end{bmatrix}$

5. $\begin{bmatrix} 3 & 2 & 6 \\ 6 & 0 & 1 \\ -2 & 10 & -11 \\ 1 & 5 & 2 \end{bmatrix}$

6. $\begin{bmatrix} 8 & 7 & 6 & 4 \\ 4 & -1 & 0 & 6 \\ -5 & 3 & -4 & 7 \end{bmatrix}$

Objective 2   **Write the augmented matrix for a system. Write the augmented matrix for each system.**

Write the augmented matrix for each system.

7. $3x - 4y = 7$
   $2x + y = 12$

8. $2x - 3y = 12$
   $7x + 3y = 15$

9. $\dfrac{1}{3}x - \dfrac{1}{2}y = 7$
   $\dfrac{5}{3}x + \dfrac{1}{2}y = 8$

10. $\dfrac{1}{2}x + \dfrac{1}{2}y = -16$
    $-3x + y = 2$

11. $3x = y + 4$
    $y = 5x - 2$

12. $y + 3 = -2x$
    $x = 4y - 5$

13. $-2x + 3y - 5z = 7$
    $6x + 2y - 4z = 12$
    $5x - 2y + z = -1$

14. $x + y + z = 10$
    $2x + y - 3z = 11$
    $x + 2z = -2$

Objective 3   **Use row operations to solve a system with two equations.**

Use row operations to solve each system.

15. $x + 3y = -7$
    $4x + 3y = -1$

16. $x - 2y = -1$
    $2x + y = 8$

17. $3x - 3y = 15$
$2x + y = 4$

18. $x - 3y = 6$
$2x + 3y = -6$

19. $3x - 2y = -6$
$2x - y = -3$

20. $2x - y = 3$
$3x + y = 2$

21. $y = x - 2$
$2x = -3y + 9$

22. $x = 3y + 4$
$2y = 5x + 19$

23. $3x + 2y = 12$
$5x - 2y = 12$

Objective 4 **Use row operations to solve a system with three equations.**

Use row operations to solve each system.

24. $x - y - z = 6$
$-x + 3y + 2z = -11$
$3x + 2y + z = 1$

25. $x + y + z = 5$
$x - 2y + 3z = 16$
$2x - y + z = 9$

26. $x + 2y + 3z = 1$
$x + y + 2z = 0$
$2x - y - z = 1$

27. $x + y + z = 6$
$x + 2y - 3z = -11$
$-2x + y - z = -11$

28. $2x + y + z = 8$
$x - y + z = 3$
$3x + y - z = 1$

29. $2x - y + 2z = -1$
$-x - 3y + z = 1$
$x + y + z = 1$

Objective 5 **Use row operations to solve special systems.**

Use row operations to solve each system.

30. $x + y = 3$
$3x + 3y = -2$

31. $x - 3y = 1$
$4x - 12y = 5$

32. $x - 2y = 3$
$3x - 6y = 9$

33. $2x + y = 10$
$-4x - 2y = -20$

34. $x + y + z = 6$
$x - y - z = 2$
$2x + y + z = 8$

35. $x - 2y + 3z = 1$
$2x - y + 3z = 2$
$4x - y + 5z = 5$

36. $x + 3y - z = 1$
$2x + y - z = 2$
$4x - 3y - z = 4$

37. $x + 2y - z = 6$
$2x + 4y - 2z = 12$
$-3x - 6y + 3z = -18$

5.4   Mixed Exercises

Give the dimensions of each matrix.

38. $\begin{bmatrix} 2 & 1 & -1 \\ 0 & 4 & 5 \end{bmatrix}$

39. $\begin{bmatrix} 4 & 8 & -10 \\ -8 & 0 & 4 \\ 5 & -1 & 2 \\ 7 & 3 & -4 \end{bmatrix}$

Write the augmented matrix for each system.

40. $2x - 5y = 2$
$3x + 4y = -1$

41. $y = -2x + 7$
$x = 4y$

Use row operations to solve each system.

42. $x - 3y = -9$
$\dfrac{1}{3}x - y = 2$

43. $x - 2y = 4$
$-2x + y = 1$

44. $x - 3y = 10$
$2x + y = 6$

45. $2x - y = 2$
$6x - 3y = 6$

46. $x - 2y + z = 4$
$3x + 3y - 3z = -1$
$4x + y - 2z = 1$

47. $x - 4y - z = 6$
$2x - y + 3z = 0$
$3x - 2y + z = 4$

48. $2x - 2y - z = -16$
$-2x + 2y + 3z = 20$
$x + y + z = 1$

Chapter 6

EXPONENTS AND POLYNOMIALS

6.1 Integer Exponents and Scientific Notation

Objective 1 Use the product rule for exponents.

Apply the product rule for exponents, if possible, in each case.

1. $3^2 \cdot 3^4$

2. $2^3 \cdot 2^5$

3. $x^3 \cdot x^5$

4. $x^8 \cdot x^7 \cdot x^4$

5. $y^3 \cdot y^6$

6. $x^4 \cdot x^2$

7. $y^3 \cdot y^8$

8. $2x^5(3x^4)$

9. $(-4p^4)(-3p^7)$

10. $(-4r^3)(3r^4)$

11. $(4z^2)(-3z^3)$

12. $(x^{14})(3x^{12})$

13. $(-9t^5)(-7t^{12})$

14. $(9y^{11})(-2y^8)$

15. $(7p^{10})(-4p^8)$

Objective 2 Define 0 and negative exponents.

Simplify.

16. 16^0

17. -38^0

18. $16x^0$

19. $-38k^2m^0$

20. $(-2p^2)(3q)^0(5r^2)$

21. $(8w^0)(-6x^2)(-2y^0)$

Simplify. Write all answers with only positive exponents. Assume all variables represent nonzero numbers.

22. 7^{-2}

23. 10^{-5}

24. $3x^{-3}$

25. $-6y^{-2}$

26. $w^4 \cdot w^{-2} \cdot w^3$

27. $p^{-6} \cdot p^0 \cdot p^3$

28. $y^{-4} \cdot y^4$

29. $3m^4 \cdot m^0 \cdot m^3 \cdot m^{-7}$

30. $(6n^3)(2n^{-5})$

31. $(-8g^7)(5g^{-2})(2g^0)$

Objective 3 Use the quotient rule for exponents.

Use the quotient rule to simplify. Write all answers with only positive exponents. Assume that variables represent nonzero real numbers.

32. $\dfrac{8^5}{8^5}$

33. $\dfrac{8^{12}}{8^{10}}$

34. $\dfrac{8^6}{8^7}$

35. $\dfrac{x^{-4}}{x^3}$

36. $\dfrac{r^4}{r^{-3}}$

37. $\dfrac{p^{-4}}{p^{-3}}$

38. $\dfrac{5^{-8}}{5^7}$

39. $\dfrac{x^4}{x^{-1}}$

40. $\dfrac{4^3}{4^{-4}}$

41. $\dfrac{5^{-7}}{5^{-8}}$

42. $\dfrac{y^{-7}}{y^2}$

43. $\dfrac{3}{3^{-1}}$

Objective 4   Use the power rules for exponents.

Use one or more power rules to simplify. Write all answers with only positive exponents. Assume all variables represent nonzero numbers.

44. $(3^2)^4$

45. $(x^2)^8$

46. $\left(\dfrac{3}{4}\right)^3$

47. $(6x)^2$

48. $(4y^4)^3$

49. $(3x^2)^4$

50. $(r^{-4})^0$

51. $(a^{-4})^{-3}$

52. $(3^{-3})^2$

53. $(3z^4)^{-2}$

54. $(3x^0)^2$

55. $(-3x^{-4})^{-3}$

56. $(6a^{-3})^4$

57. $\left(\dfrac{m^3}{3^{-4}}\right)^{-2}$

58. $\left(\dfrac{x^{-5}}{y^3}\right)^{-1}$

Objective 5   Simplify exponential expressions.

Simplify each expression. Write with only positive exponents. Assume variables represent nonzero real numbers.

59. $4^7 \cdot 4^{-4} \cdot 4^{-8}$

60. $6^2 \cdot 6^{-6} \cdot 6^{-2}$

61. $7^3 \cdot 7^{-5} \cdot 7^{-8}$

62. $6y^{-2}\left(3y^4\right)$

63. $\dfrac{9^{-3}}{9^7 \cdot 9^{-2}}$

64. $\dfrac{9^{-2}}{9^{10} \cdot 9^{-4}}$

65. $6w^5(2w^{-2})$

66. $(3a^{-2}b^{-4})^2$

67. $(5j^4k^{-2})^2$

68. $(12x^{-5}y^4)^{-2}$

69. $(2a^{-2}b)^{-1}(3ab^{-1})^{-2}$

70. $(3x^2y^{-2})^{-2}(2x^{-2}y)^{-3}$

71. $\dfrac{2^{-3}r^{-2}(r^{-1})^{-2}}{r(r^3)^{-3}}$

72. $\dfrac{3^{-2}x^{-4}(x^2)^{-3}}{7(x^2)^{-1}}$

73. $\left(\dfrac{4}{7}\right)^{-2}$

74. $\left(\dfrac{5}{6}\right)^{-1}$

75. $\left(\dfrac{3q}{4p^2}\right)^2\left(\dfrac{2p}{5q}\right)^{-2}$

76. $\dfrac{(5k^2)^3(2k^2)}{k^{-1}k^{-3}}(3k^2)^{-3}$

Objective 6   Use the rules for exponents with scientific notation.

Write each number in scientific notation.

77. 340

78. .0034

79. 14

80. .027

81. .0000382

82. 168,000

83. .168

84. .0472

85. 83,632

86. 93,000,000

87. .000000375

88. 37.5

Write each number in standard notation.

89. 3.42×10^5

92. 7.15×10^{-1}

95. 6.24×10^{-5}

90. 8.2×10^{-4}

93. 4.169×10^0

96. 9.3×10^6

91. 2.22×10^3

94. 5.83×10^4

Use scientific notation and the rules for exponents to find each value.

97. $\dfrac{8 \times 10^2}{4 \times 10^{-2}}$

100. $\dfrac{600,000}{.002}$

103. $\dfrac{7000}{.01 \times 2000}$

98. $\dfrac{12 \times 10^{-4}}{4 \times 10^{-1}}$

101. $\dfrac{.44}{880,000}$

104. $\dfrac{144,000}{.016 \times 900}$

99. $\dfrac{68,000}{.0034}$

102. $\dfrac{210,000}{30 \times .007}$

6.1 Mixed Exercises

Simplify each expression. Write with only positive exponents. Assume that variables represent nonzero real numbers.

105. $4^2 \cdot 4^{-3}$

110. $\dfrac{7^{-6}}{7^{-4}}$

113. $5x^2 \left(x^{-7}\right)\left(-4x^{-3}\right)$

106. $10^5 \cdot 10^{-2}$

107. $a \cdot a^{-1}$

111. $\dfrac{m^{-3}}{m^4}$

114. $\dfrac{12w^7 w^{-3}}{20w^{-1} w^5}$

108. $p^2 \cdot p^8 \cdot p^{-4}$

112. $\dfrac{t^2}{t^{-2}}$

109. $\dfrac{4^7}{4^{-2}}$

Simplify. Write all answers with only positive exponents. Assume variables represent nonzero real numbers.

115. $(-2y^4)^5$

119. $(3^{-1}x^{-2})^{-4}$

123. $\dfrac{(9r^2)^{-2}}{r^{-3}r^{-4}}(2r^{-2})^3$

116. $\dfrac{14^{-6}}{14^8 \cdot 14^{-12}}$

120. $\dfrac{5^{-2} y^{-4}(y^2)^{-3}}{5(y^{-2})^{-4}}$

124. $(4c^{-3}d^2)^{-3}(8c^{-4}d^{-1})^2$

117. $\left(\dfrac{m^{-4}}{p^{-3}}\right)^{-2}$

121. $\left(\dfrac{y}{2w}\right)^{-3}\left(\dfrac{w^2}{3y}\right)^2$

125. $(-5r^{-2}s^5 t^{-3})^2 (3r^2 s^{-3}t)^{-2}$

118. $15z^{-8}(4z^6)$

122. $\left(\dfrac{3}{4}\right)^{-2}$

Use scientific notation and the rules for exponents to find each value.

126. $\dfrac{2.7\times10^0}{1.8\times10^4}$

127. $\dfrac{.000081}{27,000}$

128. $\dfrac{900\times.0006}{1.8}$

129. $\dfrac{.00064}{1600\times.03}$

6.2   Adding and Subtracting Polynomials

Objective 1   Know the basic definitions for polynomials.

Give the coefficient and the degree of each term.

1. $256x^3$

2. $-12y^2z^2$

3. xy^3

4. $-m^5n$

5. $-20tx^4$

6. $67pqr$

Decide whether or not each polynomial is written in descending powers.

7. $8x^4 - 2x^3 + x^2 - x - 1$

8. $-3x^5 + 2x^2 - x^3 - 4$

9. $8 + 5y - 7y^2 + y^3$

10. $14z^6 - 3z^3 + 2z - 6$

Identify each as a trinomial, binomial, monomial, or none of these.

11. $y^2 + 3$

12. $4z^2 + 3z - 5$

13. $5a^3 + 4a^2 + 3a + 5$

14. $2x^{-3}$

15. 7

16. $3t^3 + 5t + 2$

17. $5r^5 + 4r^4$

18. $2 + s$

19. $32w^{100}$

20. $-5x + 4$

Objective 2   Find the degree of a polynomial.

Give the degree of each of these polynomials.

21. $7x^5 - 6x^3$

22. $4a$

23. -8

24. $9t^3 - 17t + 5$

25. $2n^4 + 7n^3 + 13n^{12}$

26. $p + 3p^4$

27. $-4y + 3$

28. $m^9 - 4m^{10} + 3m^2$

29. $10xy^3 - 7y^2 + xyz$

30. $6a + 3a^2$

Objective 3   Add and subtract polynomials.

Add.

31. $\begin{aligned} 14x - 8 \\ \underline{3x + 11} \end{aligned}$

32. $\begin{aligned} 5x^2 + 7x \\ \underline{-3x^2 + x - 9} \end{aligned}$

33. $\begin{aligned} -11z^2 + 3z + 1 \\ \underline{-7z^2 + 6z - 4} \end{aligned}$

34. $\begin{aligned} -4m^2 + 2m - 1 \\ \underline{3m^2 + 5m - 2} \end{aligned}$

35. $5a^3 - 4a^2 \qquad + 7$
$\underline{3a^3 + 2a^2 + 7a \qquad}$

36. $-9p^2 + 5p - 3$
$\underline{7p^2 + 7p - 4}$

37. $(3x^2 + 4x + 2) + (7x^2 - x + 2)$

38. Add $2r^3 - 6r$ and $4r - 3r^2$.

Subtract.

39. $11x + 14$
$\underline{8x + 20}$

40. $-3k + 5$
$\underline{4k - 8}$

41. $-4n^2 + 3n - 1$
$\underline{-2n^2 + 7n - 5}$

42. $5x^3 + 2x - 9$
$\underline{-3x^3 - 5x + 2}$

43. $\qquad -3y^3 + 7y^2$
$\underline{y^4 - \ y^3 \qquad + 3y}$

44. $8r^3 + 5r - 8$
$\underline{-5r^3 - 7r + 3}$

45. $(2z - 5) - (2z - 7)$

46. $(3x^4 - 7x^2 + 5) - (4x^4 + 2x^3 - 9)$

47. $6y - (3y - [3y - (6y - 9y)]) + 7y - (6y - 3y)$

6.2   Mixed Exercises

Identify each as a trinomial, binomial, monomial, or none of these. Give the degree of each.

48. $2x^2 - x - 9$

49. $5x^6 + 3x^4 - 9x + 2$

50. $9 - 3x$

51. 527

Perform the indicated operations.

52. $(4x^2 + 3x) + (8x - 9)$

53. $(-4m^3 + 8m^2 + 7m) - (m^3 + 6m^2 - 9m)$

54. $\left[-(4x^2 - 7x + 2x^3) - (2x^2 + 5x - 3x^3)\right] + x^5$

55. $(s^3 + 3s + 1) + (s^2 - 2s)$

56. Add $-4n + 2n^2$ and $2n - 4n^3$ and $4n + 5n^2$.

6.3   Polynomial Functions

Objective 1 **Recognize and evaluate polynomial functions.**

If $P(x) = 2x^2 + 3x - 5$, *find each of the following.*

1. $P(3)$ **2.** $P(-7)$ **3.** $P(0)$ **4.** $P(-1)$ **5.** $P(1)$

If $P(x) = -x^2 - x - 5$, *find each of the following.*

6. $P(-2)$ **7.** $P(3)$ **8.** $P(0)$ **9.** $P(-1)$ **10.** $P(1)$

Objective 2 **Use a polynomial function to model data.**

Solve each problem.

11. The number of medical doctors, in thousands, in the United States during the period 1990-1995 can be modeled by the polynomial function defined by
$$f(x) = 1.23x^2 + 13.9x + 616.7,$$
where $x = 0$ corresponds to 1990, $x = 1$ corresponds to 1991, and so on. Use this model to approximate the number of doctors in each given year. (*Source*: American Medical Association.)
(a) 1991 **(b)** 1992

12. The number of airports in the United States during the period from 1970 through 1997 can be approximated by the polynomial function defined by
$$f(x) = -6.77x^2 + 445.34x + 11,279.82,$$
where $x = 0$ represents 1970, $x = 1$ represents 1971, and so on. Use this function to approximate the number of airports in each given year. (*Source*: U.S. Federal Aviation Administration.)
(a) 1980 **(b)** 1992

13. The number of cases commenced by U.S. Courts of Appeals during the period from 1990 through 1998 can be approximated by the polynomial function defined by
$$f(x) = -145.32x^2 + 2610.84x + 41,341.13,$$
where $x = 0$ represents 1990, $x = 1$ represents 1991, and so on. Use this function to approximate the number of cases commenced in each given year. (*Source*: Administrative Office of the U.S. Courts, *Statistical Tables for the Federal Judiciary*, annual.)
(a) 1994 **(b)** 1998

Objective 3 **Add and subtract polynomial functions.**

For each pair of functions, find (a) $(f + g)(x)$ *and (b)* $(f - g)(x)$.

14. $f(x) = 6x - 8, g(x) = 2x + 4$ **16.** $f(x) = 2x^2 + 4x - 5, g(x) = -x^2 + 3x - 8$

15. $f(x) = -3x + 2, g(x) = 8x + 1$ **17.** $f(x) = 6x^2 - 7x + 12, g(x) = -3x^2 + x + 9$

Objective 4 **Graph basic polynomial functions.**

Graph each function by creating a table of ordered pairs. Give the domain and the range.

18. $f(x) = -3x + 2$ **19.** $f(x) = 2x - 5$ **20.** $f(x) = -2x^2$

21. $f(x) = \frac{1}{3}x^2$ **22.** $f(x) = x^3 + 2$ **23.** $f(x) = -x^3 + 1$

6.3 Mixed Exercises

If $P(x) = -x^4 + 3x^2 - x + 7$, *find each of the following.*

24. $P(-4)$ **25.** $P(3)$ **26.** $P(0)$

27. Let $f(x) = 2x^2 + 9x - 5$ and $g(x) = -3x^2 + 2x - 7$. Find each of the following.
    (a) $(f + g)(x)$ **(b)** $(f - g)(x)$

Graph each function by creating a table of ordered pairs. Give the domain and the range.

28. $f(x) = -4x^2$ **29.** $f(x) = x^3 + 3$

6.4   Multiplying Polynomials

Objective 1 **Multiply terms.**

Find each product.

1. $(4a)(5a)$

2. $(5t)(8t)$

3. $(-10s^2)(6s^3)$

4. $(5r^3)(-2r^3)$

5. $(7x^3)(-4x)$

6. $3k^4(2k^2)$

7. $(8x^2y)(9xy^3)$

8. $(-11xy^2)(2x^3y)$

9. $(3y^2z^3)(6yz^4)$

10. $(-12r^4s^7)(-9r)$

Objective 2 **Multiply any two polynomials.**

Find each product.

11. $7(3n+4)$

12. $p(3p-5)$

13. $-8y(6y-7)$

14. $2x^2(3x^2+4x+5)$

15. $-8z^3(2z^3+4z)$

16. $(4m+3)(3m^2+2m-5)$

17. $(3a-2)(a^2-a+1)$

18. $-5t^3(t^4+5t^3)$

19. $(8s+1)(3s^2+s-5)$

20. $(4x^2+2x+1)(x^2+5)$

Objective 3 **Multiply binomials.**

Use the FOIL method to find each product.

21. $(4r-2)(6r+1)$

22. $(q-5)(q+7)$

23. $(7k-4)(2k+1)$

24. $(n+4)(n-5)$

25. $(x+12)(x+7)$

26. $(x+4)(x-7)$

27. $(3z-2)(2z-1)$

28. $(5y+1)(y-3)$

29. $(2m-3)(m-4)$

30. $(6t-7)(4t+3)$

Objective 4 **Find the product of the sum and difference of two terms.**

Find each product.

31. $(5a+1)(5a-1)$

32. $(2z+3)(2z-3)$

33. $(2x-5y)(2x+5y)$

34. $(10r-s)(10r+s)$

35. $(9x+4y)(9x-4y)$

36. $(5p-3q)(5p+3q)$

37. $(8k+m)(8k-m)$

38. $\left(x+\dfrac{1}{2}\right)\left(x-\dfrac{1}{2}\right)$

39. $(4t^3+1)(4t^3-1)$

40. $(2y^2 + z^3)(2y^2 - z^3)$

Objective 5 **Find the square of a binomial.**

Find each square.

41. $(p-7)^2$

44. $(x-7y)^2$

47. $(2a-7b)^2$

49. $(2k-3m)^2$

42. $(r+5)^2$

45. $(6r-11)^2$

48. $\left(x+\dfrac{1}{2}\right)^2$

50. $(3x-4y)^2$

43. $(x+2y)^2$

46. $(5m+2n)^2$

Objective 6 **Multiply polynomial functions.**

For each pair of functions, find the product $(fg)(x)$.

51. $f(x)=3x,\ g(x)=2x-1$

54. $f(x)=x-5,\ g(x)=3x+4$

52. $f(x)=2x,\ g(x)=5x-7$

55. $f(x)=3x-8,\ g(x)=2x^2+5x+3$

53. $f(x)=x+2,\ g(x)=3x-2$

56. $f(x)=2x+5,\ g(x)=8x^2-6x+10$

6.4   Mixed Exercises

Find each product.

57. $(-3x^2y^3)(5x^3y^4)$

62. $-4q^2(-8q^4+3q-6)$

58. $(s-1)(s-3)$

63. $(3r+8)(3r-2)$

59. $(8y-3z)(3y-5z)$

64. $(3s^2+1)(3s^2-1)$

60. $(2p-5q)(2p+7q)$

65. $(x+y)(2x-3y)$

61. $(7x+10)(7x-10)$

66. $(z+10)^2$

For each pair of functions, find the product $(fg)(x)$.

67. $f(x)=x-3,\ g(x)=8x+5$

68. $f(x)=2x+1,\ g(x)=3x^2-x+9$

6.5   Dividing Polynomials

Objective 1   Divide a polynomial by a monomial.

Divide.

1. $\dfrac{8z^5}{2z}$

2. $\dfrac{36y^8}{4y^3}$

3. $\dfrac{18m^5n^6}{3m^2n^2}$

4. $\dfrac{28t^7 - 12t^2}{4t}$

5. $\dfrac{10x^5 - 16x^2 + 8x^3}{x^2}$

6. $\dfrac{8a^5 - 4a^3 + 4a^2}{4a^3}$

7. $\dfrac{15r^2 - 9r^3}{3r^2}$

8. $\dfrac{45q^3 + 15q^2 + 9q^5}{5q^3}$

9. $\dfrac{-12x^6 + 6x^5 + 3x^4 - 9x^3 + 3x}{3x}$

10. $\dfrac{7b^2 - 9b + 5}{5b}$

11. $\dfrac{5 + z + 6z^2 + 8z^3}{3z^4}$

12. $\dfrac{120a^{11} - 60a^{10} + 140a^9}{10a^{12}}$

Objective 2   Divide a polynomial by a polynomial of two or more terms.

Divide.

13. $\dfrac{m^2 - m - 2}{m - 2}$

14. $\dfrac{x^2 + 5x + 6}{x + 2}$

15. $\dfrac{2z^2 - 5z - 3}{2z + 1}$

16. $\dfrac{9x^2 + 6x - 8}{3x - 2}$

17. $\dfrac{4s^2 + 11s - 8}{s + 3}$

18. $\dfrac{12r^3 - 11r^2 + 9r + 18}{4r + 3}$

19. $\dfrac{12x^3 - 17x^2 + 30x - 10}{3x^2 - 2x + 5}$

20. $\dfrac{y^4 - 3y^2 + 7}{y^2 - 1}$

21. $\dfrac{2p^5 + 6p^4 - p^3 + 3p^2 - p}{2p^2 + 1}$

22. $\dfrac{8m^5 - 4m^4 - 2m^3 + 7m^2 - 3m + 8}{4m^2 - 3}$

Objective 3 Divide polynomial functions.

For each pair of functions, find the quotient $\left(\frac{f}{g}\right)(x)$ and give any x-values that are not in the domain of the quotient function.

23. $f(x)=15x^2-3x,\ g(x)=3x$

24. $f(x)=20x^2-35x,\ g(x)=5x$

25. $f(x)=2x^2-x-10,\ g(x)=x+2$

26. $f(x)=4x^2-11x-45,\ g(x)=x-5$

27. $f(x)=27x^3-8,\ g(x)=3x-2$

6.5 Mixed Exercises

Divide.

28. $\dfrac{6x^3-8x+9}{9x^3}$

29. $\dfrac{4q^2-4q+5}{2q-1}$

30. $\dfrac{4k^2-4k+7}{2k-1}$

31. $\dfrac{2s^2-2s+5}{s}$

32. $\dfrac{x^2+11x+24}{x+8}$

33. $\dfrac{t^2-3t-10}{t-5}$

34. $\dfrac{27p^4-36p^3-6p^2+26p-2}{3p}$

35. $\dfrac{x^4-7}{x^2-2}$

36. $\dfrac{n^5-2}{n^2-1}$

37. $\dfrac{2n^5-6n^4+8n^2}{-2n^3}$

For each pair of functions, find the quotient $\left(\frac{f}{g}\right)(x)$ and give any x-values that are not in the domain of the quotient function.

38. $f(x)=18x^3-3x^2,\ g(x)=3x^2$

39. $f(x)=2x^2-5x-42,\ g(x)=x-6$

Chapter 7

FACTORING

7.1 Greatest Common Factors; Factoring by Grouping

Objective 1 Factor out the greatest common factor.

Factor out the greatest common factor if possible.

1. $7y + 21$

3. $15x - 15y$

5. $12z - 17$

7. $12a^3 + 6a^2$

2. $32x - 8$

4. $12a - 3b$

6. $26x + 62y$

8. $48r^2 + 16r^5$

Factor, and then simplify.

9. $(p-9)(p+2) + (p-9)(p+1)$

11. $(2q+1)(q-6) - (2q+1)(q+5)$

10. $(k+5)(k-6) - (k-6)(k+1)$

12. $(x+2)(2x+3) - (x+2)(x+1)$

Objective 2 Factor by grouping.

Factor by grouping.

13. $x + y + 6ax + 6ay$

18. $x - 8y^2 + 2xy^2 - 4$

14. $x^2 - xy + 9x - 9y$

19. $8 - 12p - 6p^3 + 9p^4$

15. $3b + 3c + ab + ac$

20. $x^3 + x^3y^2 - 3y^2 - 3$

16. $4ax + 4ay + 3bx + 3by$

21. $-3x - 6 + 2y + xy$

17. $1 - x + xy - y$

22. $bx - by - ay + ax$

7.1 Mixed Exercises

Factor out the greatest common factor if possible.

23, $s^5 + 4s^6 + 8s^3$

25. $5x^3y^2 + 25x^2y^3$

27. $2x^2y^8 + 5p^3q$

24. $8m^5 + 6m^2 - 12$

26. $14x^3y^2 + 7x^2y - 21x^5y^3$

28. $6a^2b^3 + 25a^4b^5$

Factor by grouping.

29. $5rs - 5rt - 2qs + 2qt$

31. $3x^3 + 3xy^2 + 4x^2y + 4y^3$

30. $14w^2 + 6wx - 35wx - 15x^2$

32. $16x^3 - 4x^2y^2 - 4xy + y^3$

7.2   Factoring Trinomials

Objective 1   Factor trinomials when the coefficient of the squared term is 1.

Factor.

1. $z^2 + 5z + 6$

2. $y^2 + 2y - 3$

3. $x^2 + 3x - 28$

4. $m^2 - m - 12$

5. $t^2 - t - 20$

6. $x^2 + x - 30$

7. $r^2 - 6r - 16$

8. $q^2 + 6q + 5$

9. $k^2 + 9k + 20$

10. $n^2 + 6n + 9$

11. $x^2 + 8xy + 15y^2$

12. $a^2 - 2a - 35$

13. $r^2s^2 + 4rs - 21$

14. $x^2y^2 - 12xy + 27$

Objective 2   Factor trinomials when the coefficient of the squared term is not 1.

Objective 3   Use an alternative method for factoring trinomials.

Factor.

15. $3y^2 + 13y + 4$

16. $4x^2 + 5x + 1$

17. $3z^2 + 2z - 8$

18. $6t^2 + t - 1$

19. $20r^2 - 28r - 3$

20. $5x^2 + 13x + 6$

21. $20x^2 + 39x - 11$

22. $15q^2 - 7q - 4$

23. $8x^2 + 2x - 15$

24. $6p^2 - p - 15$

25. $6y^2 - 5yz - 6z^2$

26. $6x^2 - 5xy - y^2$

27. $4k^2 + 13kp + 3p^2$

28. $6x^2 + 7xy - 20y^2$

Objective 4   Factor by substitution.

Factor completely.

29. $2(x+y)^2 + 7(x+y) + 3$

30. $3(p-q)^2 + 10(p-q) + 7$

31. $5(a-1)^2 - 7(a-1) - 6$

32. $8(p+5)^2 + 2(p+5) - 15$

33. $8(5-z)^2 - 14(5-z) + 3$

34. $10(r+s)^2 - (r+s) - 24$

35. $40y^4 + y^2 - 6$

36. $18x^4 - x^2 - 5$

37. $4t^4 + 69t^2 + 17$

38. $6x^4y^4 - 7x^2y^2 - 5$

7.2   Mixed Exercises

Factor.

39. $x^2 - 6x - 27$

40. $2a^2 - 17a + 30$

41. $6m^2 - 13mn - 5n^2$

42. $x^2 - 2xy - 15y^2$

43. $4y^4 + 8y^2 - 45$

44. $25x^2 - 5xy - 2y^2$

45. $p^2 - 4p - 21$

46. $3(a+1)^2 + 19(a+1) - 14$

47. $12m^2 + 11m - 5$

48. $p^2 + 12p + 32$

7.3 Special Factoring

Objective 1 Factor a difference of squares.

Factor.

1. $y^2 - 16$

2. $9x^2 - 1$

3. $36z^2 - 121$

4. $x^2 + 49$

5. $144x^2 - 25y^2$

6. $a^2 - 100$

7. $s^4 - 16$

8. $x^2 - 81$

9. $(x+y)^2 - 25$

10. $q^2 - (2r+3)^2$

11. $25 - (x-y)^2$

12. $(r-s)^2 - (r+s)^2$

Objective 2 Factor a perfect square trinomial.

Factor each perfect square trinomial.

13. $x^2 + 4x + 4$

14. $z^2 - 10z + 25$

15. $x^2 + 8x + 16$

16. $y^2 + 22y + 121$

17. $x^2 + 24x + 144$

18. $k^2 + 12k + 36$

19. $16a^2 - 40ab + 25b^2$

20. $16t^2 + 56t + 49$

21. $36r^2 - 60rs + 25s^2$

22. $25x^2 - 20xy + 4y^2$

23. $(y+z)^2 + 14(y+z) + 49$

24. $(p-q)^2 - 20(p-q) + 100$

25. $4x^2 - 4xy + y^2$

26. $(m-n)^2 - 12(m-n) + 36$

Objective 3 Factor a difference of cubes.

Factor.

27. $x^3 - y^3$

28. $8a^3 - 1$

29. $8r^3 - 27s^3$

30. $64x^3 - y^3$

31. $216m^3 - 125p^6$

32. $8a^3 - 125b^3$

33. $(r+s)^3 - 1$

34. $(m+n)^3 - (m-n)^3$

35. $x^3 - (x-1)^3$

36. $216x^3 - 8y^3$

Objective 4 Factor a sum of cubes.

Factor.

37. $x^3 + y^3$

38. $z^3 + 8$

39. $27r^3 + 8s^3$

40. $8a^3 + 64b^3$

41. $125p^3 + q^3$

42. $64x^3 + 343y^3$

43. $1 + (y + z)^3$

45. $(a - 1)^3 + a^3$

44. $(x - y)^3 + (x + y)^3$

46. $t^3 + (t + 2)^3$

7.3   Mixed Exercises

Factor.

47. $x^2 - 14x + 49$

48. $x^4 - 81$

49. $125m^3 + 8n^3$

50. $16 - (p - q)^2$

51. $(m - p)^2 - (m - p)^2$

52. $y^6 + 1$

53. $125m^3 - 64p^3$

54. $4x^2 + 12xy + 9y^2$

55. $y^2 - z^2$

56. $z^3 - 125y^3$

7.4   Solving Equations by Factoring

Objective 1   Learn and use the zero-factor property.

Solve each equation using the zero-factor property.

1. $(2x+3)(x-4)=0$

2. $(r+4)(r-6)=0$

3. $(4y+3)(y-12)=0$

4. $8x^2-24x=0$

5. $r^2+r-72=0$

6. $p^2+p-20=0$

7. $6z^2+19z+10=0$

8. $8x^2=15-2x$

9. $12a^2+11a=5$

10. $5x^2-4=8x$

11. $6r^2=-3-11r$

12. $5p^2=-3p$

13. $2k^2+3k=9$

14. $25x^2-36=0$

Objective 2   Solve applied problems that require the zero-factor property.

Solve each problem by writing a quadratic equation and then solving it using the zero-factor property.

15. The sum of two numbers is 7. The sum of their squares is 25. Find the numbers.

16. Two numbers have a sum of 12. The sum of the squares of the numbers is 144. Find the numbers.

17. Find two consecutive whole numbers whose product is 156.

18. Find two consecutive even whole numbers whose product is 168.

19. Find two consecutive odd whole numbers whose product is 143.

20. Find two consecutive even whole numbers whose product is 288.

21. Two numbers have a sum of 3 and a product of -70. Find the numbers.

22. Two numbers have a sum of -6 and a product of -16. Find the numbers.

23. The top of a table has an area of 63 square feet. It has a length that is 2 feet more than the width. Find the width of the table top.

24. A house has a floor area of 608 square meters. The floor has the shape of a rectangle whose length is 13 meters more than the width. Find the width of the floor.

7.4   Mixed Exercises

Solve each equation using the zero-factor property.

25. $4q^2-5q-6=0$

26. $16x^2-25=0$

27. $16m^2-64=0$

28. $z^2=6z-9$

29. $15s^2-2=s$

30. $3x^2=5x+28$

Solve each problem by writing a quadratic equation and then solving it using the zero-factor property.

31. The sum of two numbers is 4. The sum of their squares is 136. Find the numbers.

32. Paul and Joan Rice wish to buy floor covering that covers 150 square feet for their large recreation room. They wish to cover a rectangle that is 5 feet longer than it is wide. How wide should the rectangle be?

33. Mary Ann Young has enough grass seed to cover 400 square feet. She wishes to grow grass on a rectangular area that is 9 feet longer than it is wide. What dimensions should the area have?

34. The Browns installed 96 feet of fencing around a rectangular play yard. If the yard covers 540 square feet, what are its dimensions?

Chapter 8

RATIONAL EXPRESSIONS AND FUNCTIONS

8.1 Rational Expressions and Functions; Multiplying and Dividing

Objective 1 Define rational expressions.

Objective 2 Define rational functions and describe their domains.

Find all numbers that are not in the domain of the function.

1. $f(x) = \dfrac{4}{x-5}$

2. $f(x) = \dfrac{x}{x+5}$

3. $f(x) = \dfrac{2x+3}{x+7}$

4. $f(m) = \dfrac{m+7}{m}$

5. $f(t) = \dfrac{t-9}{4}$

6. $f(a) = \dfrac{2a-3}{4a-7}$

7. $f(s) = \dfrac{8s+7}{3s-2}$

8. $f(r) = \dfrac{r+7}{r^2-25}$

9. $f(q) = \dfrac{q+7}{q^2-3q+2}$

10. $f(x) = \dfrac{2x-5}{x^2-10x+25}$

11. $f(x) = \dfrac{x-6}{x^2+1}$

12. $f(x) = \dfrac{x^2-4}{x^2+4}$

Objective 3 Write rational expressions in lowest terms.

Write in lowest terms.

13. $\dfrac{4n^5}{16n^3}$

14. $\dfrac{35p^5}{15}$

15. $\dfrac{-6x^3y^7}{18xy^4}$

16. $\dfrac{(m-7)(m+3)}{(m+3)(m-2)}$

17. $\dfrac{k(k+4)}{5k(k+4)}$

18. $\dfrac{3x-3}{4x-4}$

19. $\dfrac{6y^2+y}{3y^2+y}$

20. $\dfrac{x^2-6x+9}{x^2-9}$

21. $\dfrac{11r^2-22r^3}{6-12r}$

22. $\dfrac{s^2-s-6}{s^2+s-12}$

23. $\dfrac{8z^2+6z-9}{16z^2-9}$

24. $\dfrac{-x+y}{y-x}$

25. $\dfrac{c-2d}{2d-c}$

26. $\dfrac{x^2-1}{1-x}$

27. $\dfrac{a^2-3a}{3a-a^2}$

28. $\dfrac{r^3-s^3}{r^2-s^2}$

29. $\dfrac{x^2-y^2}{x^3-y^3}$

30. $\dfrac{(p-5)(5-p)}{(5-p)(5+p)}$

| Objective 4 | Multiply rational expressions. |

Multiply.

31. $\dfrac{9x^2}{16} \cdot \dfrac{4}{3x}$

37. $\dfrac{3(s-1)}{s} \cdot \dfrac{2s}{5(s-1)}$

32. $\dfrac{21z^4}{8z} \cdot \dfrac{4z^3}{7z^5}$

38. $\dfrac{5t+25}{10} \cdot \dfrac{12}{6t+30}$

33. $\dfrac{4r^2}{8r} \cdot \dfrac{3r^3}{4r^4}$

39. $\dfrac{9k-18}{6k+12} \cdot \dfrac{3k+6}{15k-30}$

34. $\dfrac{6y^3}{9y} \cdot \dfrac{12y}{y^2}$

40. $\dfrac{x^2-6x}{9x} \cdot \dfrac{18x}{3x-18}$

35. $\dfrac{10a^3}{20a^2} \cdot \dfrac{12a^4}{3a}$

41. $\dfrac{x^2-x-6}{x^2-2x-8} \cdot \dfrac{x^2+7x+12}{x^2-9}$

36. $\dfrac{m+3}{2} \cdot \dfrac{12}{(m+3)^2}$

42. $\dfrac{6z^2-5z-6}{6z^2+5z-6} \cdot \dfrac{12z^2-17z+6}{12z^2-z-6}$

| Objective 5 | Find reciprocals for rational expressions. |

Find the reciprocal.

43. $\dfrac{3}{y}$

46. $\dfrac{m^2+2m+3}{5}$

49. $\dfrac{8-s}{s-8}$

52. $\dfrac{7z+7}{z^2-9}$

44. $\dfrac{4}{p-5}$

47. $\dfrac{2p-1}{p^2+7p}$

50. $\dfrac{r^2+2r}{5+r}$

53. 0

45. $\dfrac{x^2+9}{7x}$

48. $\dfrac{n+8}{7}$

51. $\dfrac{x^2+4}{3x-6}$

54. $\dfrac{x^2-3x+4}{x^2+x+2}$

| Objective 6 | Divide rational expressions. |

Divide.

55. $\dfrac{7x^4}{6x^2} \div \dfrac{14x^3}{3x}$

58. $\dfrac{5r^3}{4r^2} \div \dfrac{15r^2}{8r^4}$

56. $\dfrac{15t^{10}}{9t^5} \div \dfrac{6t^6}{10t^4}$

59. $\dfrac{k-3}{16} \div \dfrac{k-3}{32}$

57. $\dfrac{3s^4}{4s^3} \div \dfrac{9s^3}{32s^4}$

60. $\dfrac{12}{4x-12} \div \dfrac{2}{x-3}$

61. $\dfrac{2n+8}{6} \div \dfrac{3n+12}{2}$

62. $\dfrac{12p+24}{36p-36} \div \dfrac{6p+12}{8p-8}$

63. $\dfrac{a^2-16}{a+3} \div \dfrac{a-4}{a^2-9}$

64. $\dfrac{6(m+2)}{3(m-1)^2} \div \dfrac{(m+2)^2}{9(m-1)}$

65. $\dfrac{4z+12}{2z-10} \div \dfrac{z^2-9}{z^2-z-20}$

66. $\dfrac{8-y}{y-8} \div \dfrac{y-8}{y+8}$

67. $\dfrac{x^2-x-6}{x^2+x-12} \div \dfrac{x^2+2x-3}{x^2+3x-4}$

68. $\dfrac{16-r^2}{r^2+2r-8} \div \dfrac{r^2-2r-8}{4-r^2}$

69. $\dfrac{s^2-s-2}{s^2+5s+4} \div \dfrac{s-2}{s+3}$

70. $\dfrac{2a^2-5a-12}{a^2-10a+24} \div \dfrac{4a^2-9}{a^2-9a+18}$

8.1   Mixed Exercises

Write each rational expression in lowest terms, and multiply or divide as indicated.

71. $\dfrac{6q^2r}{30qr^3}$

72. $\dfrac{6-3x}{2x-4}$

73. $\dfrac{z^2-9z}{9-z}$

74. $\dfrac{8r+r^2}{r+8}$

75. $\dfrac{2-y}{8} \cdot \dfrac{7}{y-2}$

76. $\dfrac{5-x}{4} \cdot \dfrac{12}{x-5}$

77. $\dfrac{(t+1)^3(t+4)}{t^2+5t+4} \div \dfrac{t^2+2t+1}{t^2+3t+2}$

78. $\dfrac{x^2-25}{5+x} \cdot \dfrac{x}{5-x}$

79. $\dfrac{p^2+3p+2}{p^2-3p-4} \cdot \dfrac{p^2-10p+24}{p^2+5p+6}$

80. $\dfrac{(z+4)^2(z-1)}{z^2+3z-4} \div \dfrac{z^2-16z}{z^2+8z+16}$

8.2 Adding and Subtracting Rational Expressions

Objective 1 Add and subtract rational expressions with the same denominator.

Add or subtract as indicated. Write all answers in lowest terms.

1. $\dfrac{3}{x} + \dfrac{8}{x}$

2. $\dfrac{5}{y^2} - \dfrac{8}{y^2}$

3. $\dfrac{3}{5t} + \dfrac{15}{5t}$

4. $\dfrac{c}{5a} - \dfrac{4}{5a}$

5. $\dfrac{n}{m+3} - \dfrac{-3n+7}{m+3}$

6. $\dfrac{z^2}{z-y} - \dfrac{y^2}{z-y}$

7. $\dfrac{r}{r^2-s^2} + \dfrac{s}{r^2-s^2}$

8. $\dfrac{4x+3}{x-7} - \dfrac{3x+10}{x-7}$

9. $\dfrac{x}{x^2-7x+10} - \dfrac{2}{x^2-7x+10}$

10. $\dfrac{k}{k^2+3k-10} - \dfrac{2}{k^2+3k-10}$

11. $\dfrac{1}{q^2-6q-7} + \dfrac{q}{q^2-6q-7}$

12. $\dfrac{b}{a^2-b^2} - \dfrac{a}{a^2-b^2}$

Objective 2 Find a least common denominator.

Assume that the expressions given are denominators of fractions. Find the least common denominator (LCD) for each group.

13. $5m, 6m$

14. $25z, 30z$

15. $5x, 15x^2, 25xy$

16. $t, t-1$

17. $8s + 24, 3s + 9$

18. $5a + 10, a^2 + 2a$

19. $q^2 - 36, (q+6)^2$

20. $r - p, p - r$

21. $r^2 + 5r + 4, r^2 + r$

22. $3n + n^2, 3 - n$

23. $p - 4, p^2 - 16, (p+4)^2$

24. $2z^2 + 7z - 4, 2z^2 - 7z + 3$

Objective 3 Add and subtract rational expressions with different denominators.

Add or subtract as indicated. Write all answers in lowest terms.

25. $\dfrac{5}{y} + \dfrac{4}{7}$

26. $\dfrac{9}{x} + \dfrac{3}{2}$

27. $\dfrac{3}{5} - \dfrac{1}{z}$

28. $\dfrac{5a}{6} - \left(\dfrac{2a}{3} - \dfrac{a}{6} \right)$

29. $\dfrac{4+2m}{5}+\dfrac{2+m}{10}$

30. $\dfrac{6}{s^2}-\dfrac{2}{s}$

31. $\dfrac{6}{5t}+\dfrac{4}{3t^2}$

32. $\dfrac{3r+4}{3}+\dfrac{6r+4}{6}$

33. $\dfrac{1}{x^2-1}-\dfrac{1}{x^2+3x+2}$

34. $\dfrac{y+9}{y^2-16}+\dfrac{2}{y+4}$

35. $\dfrac{4}{2-a}+\dfrac{7}{a-2}$

36. $\dfrac{-1}{3-m}-\dfrac{2}{m-3}$

37. $\dfrac{5r}{r+2s}-\dfrac{3r}{-r-2s}$

38. $\dfrac{a+3b}{b^2+2ab+a^2}+\dfrac{a-b}{3b^2+4ab+a^2}$

8.2 Mixed Exercises

Find the least common denominator for each group of rational expressions.

39. $\dfrac{-7}{2-m},\ \dfrac{3}{m-2}$

40. $\dfrac{x+5}{x^2+5x+6},\ \dfrac{3-x}{3x+6}$

Add or subtract as indicated. Write all answers in lowest terms.

41. $\dfrac{2x-1}{x^2+x-2}-\dfrac{x}{x^2+x-2}$

42. $\dfrac{1}{r-6}-\dfrac{2}{6-r}$

43. $\dfrac{4}{3a}+\dfrac{7}{2a^2b}$

44. $\dfrac{q+2}{q}+\dfrac{q}{q+2}$

45. $\dfrac{8}{k-2}-\dfrac{4}{k+2}$

46. $\dfrac{6}{m+n}+\dfrac{2}{m-n}$

47. $\dfrac{z}{z^2-1}+\dfrac{z-1}{z^2+2z+1}$

48. $\dfrac{2}{4y^2-16}+\dfrac{3}{4+2y}$

8.3 Complex Fractions

Objective 1 Simplify complex fractions by simplifying the numerator and denominator. (Method 1)

Use Method 1 to simplify.

1. $\dfrac{\frac{x+1}{x}}{\frac{x+1}{y}}$

2. $\dfrac{\frac{2p+3}{5}}{\frac{8p+12}{3}}$

3. $\dfrac{\frac{3}{k}+1}{\frac{3+k}{2}}$

4. $\dfrac{\frac{1}{t}+\frac{1}{z}}{\frac{1}{z+t}}$

5. $\dfrac{\frac{1}{s}+r}{\frac{1}{r}+s}$

6. $\dfrac{\frac{m+1}{m-1}}{\frac{1}{m+1}}$

7. $\dfrac{q+\frac{1}{q+1}}{q-\frac{1}{q}}$

8. $\dfrac{\frac{1}{a+b}}{\frac{4}{a^2-b^2}}$

9. $\dfrac{\frac{rs}{3r^2}}{\frac{s^2}{3}}$

Objective 2 Simplify complex fractions by multiplying by a common denominator. (Method 2)

Use Method 2 to simplify.

10. $\dfrac{\frac{3}{k}}{\frac{9}{k^2}}$

11. $\dfrac{\frac{6}{m}}{\frac{12}{m^2}}$

12. $\dfrac{x+\frac{2}{x}}{\frac{x^2+2}{3}}$

13. $\dfrac{\frac{a-2b}{a}}{\frac{a-2b}{b}}$

14. $\dfrac{\frac{s}{s+1}}{\frac{5}{2(s+1)}}$

15. $\dfrac{p+\frac{1}{p}}{\frac{3}{p}-p}$

16. $\dfrac{t-\frac{2}{t}}{t+\frac{4}{t}}$

17. $\dfrac{\frac{2}{x-1}+2}{\frac{2}{x+1}-2}$

18. $\dfrac{\frac{r}{r+1}+1}{\frac{2r+1}{r-1}}$

Objective 3 Compare the two methods of simplifying complex fractions.

Use either method to simplify each complex fraction.

19. $\dfrac{\frac{m+n}{m}}{\frac{1}{m}+\frac{1}{n}}$

20. $\dfrac{a+\frac{1}{a}}{\frac{1}{a}-a}$

21. $\dfrac{\frac{z-y}{z+y}}{\frac{z}{z-y}}$

22. $\dfrac{\frac{1}{t+1}-1}{\frac{1}{t-1}+1}$

23. $\dfrac{\frac{25k^2-m^2}{4k}}{\frac{5k+m}{7k}}$

24. $\dfrac{\frac{1}{r}}{\frac{1+r}{1-r}}$

Objective 4 Simplify rational expressions with negative exponents.

Simplify each expression, using only positive exponents in the answer.

25. $\dfrac{x^{-1}}{x^{-1}+5}$

26. $\dfrac{1+x^{-1}}{(x+y)^{-1}}$

27. $x^{-1}+y^{-2}$

28. $\dfrac{2x^{-1}+y^2}{z^{-3}}$

29. $\dfrac{z^{-2}}{4+y^{-3}}$

30. $(x^{-2}-y^{-2})^{-1}$

31. $\dfrac{4x^{-2}}{2+6y^{-3}}$

33. $\dfrac{x^{-1}}{y-x^{-1}}$

32. $\dfrac{s^{-1}+r}{r^{-1}+s}$

34. $\dfrac{(m+n)^{-2}}{m^{-2}-n^{-2}}$

8.3 Mixed Exercises

Use either method to simplify each complex fraction.

35. $\dfrac{\frac{1}{x}+\frac{1}{x-1}}{\frac{1}{x}-\frac{2}{x-1}}$

37. $\dfrac{\frac{m}{n+1}}{\frac{m^2}{n}}$

39. $\dfrac{\frac{m}{4}-\frac{1}{m}}{1+\frac{m+4}{m}}$

36. $\dfrac{\frac{5p}{y^3}}{\frac{5p^2}{y^3}}$

38. $\dfrac{n-\frac{n+2}{4}}{\frac{3}{4}-\frac{5}{2n}}$

40. $\dfrac{\frac{s}{s+1}}{\frac{2}{s^2-1}}$

Simplify each expression, using only positive exponents in the answer.

41. $\dfrac{x^{-1}-y^{-1}}{x^{-1}+y^{-1}}$

42. $(x^{-4}+4)^{-1}$

43. $\dfrac{x^{-1}-x}{x^{-1}+1}$

44. $\dfrac{x^{-2}-y^{-2}}{x+y}$

8.4   Equations with Rational Expressions and Graphs

Objective 1   Determine the domain of a rational equation.

(a) Without actually solving the equations below, list all possible numbers that would have to be rejected if they appeared as potential solutions. (b) Then give the domain using set notation.

1. $\dfrac{1}{x} + \dfrac{2}{x+1} = 0$

2. $\dfrac{4}{2x-5} - \dfrac{6}{3x+1} = \dfrac{1}{2}$

3. $\dfrac{10}{x-7} + \dfrac{7}{8+x} = 0$

4. $\dfrac{x}{6} - \dfrac{1}{2} = \dfrac{3}{x+1}$

5. $\dfrac{4}{x^2-2x} + \dfrac{x}{2} = 0$

6. $\dfrac{3}{3+x} + \dfrac{5}{x^2-x} = \dfrac{9}{x}$

Objective 2   Solve rational equations.

Solve each equation.

7. $\dfrac{6}{x} = 4 - \dfrac{5}{x}$

8. $\dfrac{2p}{7} - 5 = p$

9. $\dfrac{z+1}{2} = \dfrac{z+2}{3}$

10. $\dfrac{9}{m-2} = 3$

11. $\dfrac{2y+3}{y} = \dfrac{3}{2}$

12. $\dfrac{5-a}{a} + \dfrac{3}{4} = \dfrac{7}{a}$

13. $\dfrac{5x-3}{4x+7} = \dfrac{7}{15}$

14. $\dfrac{2}{t} = \dfrac{t}{5t-12}$

15. $\dfrac{x}{2x+2} = \dfrac{-2x}{4x+4} + \dfrac{2x-3}{x+1}$

16. $\dfrac{4}{n} - \dfrac{2}{n+1} = 3$

| Objective 3 | Recognize the graph of a rational function.

Graph each rational function. Give the equation of the vertical asymptote.

17. $f(x) = \dfrac{5}{x}$

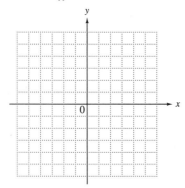

19. $f(x) = \dfrac{2}{x-1}$

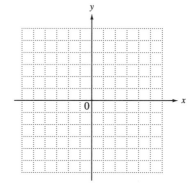

18. $f(x) = \dfrac{1}{x-3}$

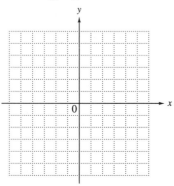

20. $f(x) = \dfrac{1}{x-4}$

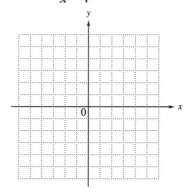

8.4   Mixed Exercises

Solve each equation.

21. $\dfrac{2w+1}{5} = \dfrac{w+1}{4}$

22. $\dfrac{x}{8} + \dfrac{x}{4} = 8$

23. $\dfrac{3}{4p} - \dfrac{2}{p} = \dfrac{5}{12}$

24. $\dfrac{-3}{r} = 2 + \dfrac{1}{r}$

25. $\dfrac{x+3}{x} - \dfrac{x+4}{x+5} = \dfrac{15}{x^2+5x}$

26. $\dfrac{y}{y-4} + \dfrac{2}{y} = \dfrac{16}{y^2-4y}$

27. $\dfrac{q+12}{q^2-16} - \dfrac{3}{q-4} = \dfrac{1}{q+4}$

28. $x - 11 + \dfrac{18}{x} = 0$

29. $\dfrac{3p}{p^2+5p+6} = \dfrac{5p}{p^2+2p-3} - \dfrac{2}{p^2+p-2}$

30. $\dfrac{1}{b+2} - \dfrac{5}{b^2+9b+14} = \dfrac{-3}{b+7}$

8.5   Applications of Rational Expressions

Objective 1   Find the value of an unknown variable in a formula.

Find the value of the variable indicated.

1. If $m = \frac{a}{b}$, $m = 9$, and $a = 5$, find b.

2. If $I = \frac{E}{R}$, $I = 12$, and $E = 4$, find R.

3. If $F = \frac{GmM}{d^2}$, $F = 150$, $G = 32$, $M = 50$, and $d = 10$, find m.

4. If $c = \frac{100b}{L}$, $c = 80$, and $b = 16$, find L.

5. If $\frac{1}{R} = \frac{1}{R_1} + \frac{1}{R_2}$, $R = 10$, and $R_1 = 20$, find R_2.

6. If $\frac{1}{f} = \frac{1}{d_0} + \frac{1}{d_i}$, $f = 10$, and $d_0 = 25$, find d_i.

7. If $r = \frac{d}{t}$, $r = 60$, and $d = 120$, find t.

8. If $t = \frac{I}{pr}$, $t = 3$, $p = 100$, and $I = 15$, find r.

9. If $h = \frac{2A}{B+b}$, $A = 40$, $h = 8$, and $b = 3$, find B.

10. If $B = \frac{3V}{h}$, $B = 20$, and $V = 80$, find h.

Objective 2   Solve a formula for a specified variable.

Solve each formula for the specified variable.

11. $\dfrac{1}{R} = \dfrac{1}{R_1} + \dfrac{1}{R_2}$ for R_2

12. $\dfrac{1}{f} = \dfrac{1}{d_0} + \dfrac{1}{d_i}$ for d_0

13. $\dfrac{V_1 P_1}{T_1} = \dfrac{V_2 P_2}{T_2}$ for T_2

14. $F = \dfrac{Gm_1 m_2}{d^2}$ for m_2

15. $s_n = \dfrac{n}{2}(a_1 + a_n)$ for a_n

16. $A = \dfrac{1}{2}h(b_1 + b_2)$ for b_1

17. $\dfrac{V_1 P_1}{T_1} = \dfrac{V_2 P_2}{T_2}$ for V_1

18. $A = \dfrac{R_1 R_2}{R_1 + R_2}$ for R_1

19. $I = \dfrac{nE}{R + nr}$ for R

20. $E = \dfrac{e(R + r)}{r}$ for r

Objective 3   Solve applications using proportions.

Use proportions to solve each problem.

21. A student is expected to answer 5 of 6 questions correctly on a certain pretest. If there are 24 questions on the test, how many questions is a student expected to answer correctly?

22. In a certain midwestern city in a recent year, there were 500 crimes committed per 100,000 population. If the population of that city was 350,000, how many crimes were committed?

23. If Jennifer can address her wedding invitations in $4\frac{1}{2}$ hours, what is her rate (in job per hour)?

24. Alex paid $1.24 in sales tax on a purchase of $15.50. How much sales tax would he pay on a purchase of $480 at the same tax rate?

25. Steven invested $20,000 and earned $900 in income in the first year. How much income would he have earned if he had invested $25,000 in the same account?

26. Holly drove 315 miles in 7 hours. How long would it take her to drive 225 miles if she traveled at the same rate?

27. Ryan's car travels 24 miles using 1 gallon of gas. If Ryan has 4 gallons of gas in his car, how much more gas will he need to travel 288 miles?

28. Connie paid $.90 in sales tax on a purchase of $12.00. Later that day she purchased an item for which she paid $55.90 including sales tax. How much was the sales tax on that item?

29. A certain city with a population of 400,000 had 1600 burglaries during the last year. If the number of burglaries increased by 200 this year and its burglary rate remained the same, by what number did its population increase?

30. Megan has invested $5000 in an account that earns $200 in income per year. If she wishes to earn $240 per year at the same rate, by how much must she increase her investment.

Objective 4   Solve applications about distance, rate, and time.

Solve each problem.

31. Kate averages 10 miles per hour riding her bike to town. Averaging 30 miles per hour by car takes her 2 hours less time. How far does she travel to town?

32. Emily's boat goes 14 miles per hour. Find the speed of the current in the river if she can go 8 miles downstream in the same time as she can go 6 miles upstream.

33. Lauren's boat can go 9 miles per hour in still water. How far downstream can Lauren go if the river has a current of 3 miles per hour and she must be back in 4 hours.

34. A canal has a current of 2 miles per hour. Find the speed of Leroy's boat in still water if it goes 30 miles downstream in the same time as it takes to go 18 miles upstream.

35. Wilmoth can fly his plane 180 miles against the wind in the same time it takes him to fly 540 miles with the wind. The wind blows at 30 miles per hour. Find the speed of his plane in still air.

36. Pauline and Pete agree to meet in Columbia. Pauline travels 120 miles, while Pete travels 80 miles. If Pauline's speed is 20 miles per hour greater than Pete's and they both spend the same amount of time traveling, at what speed does each travel?

37. Tim traveled from here to Bristol at 20 miles per hour and from Bristol to here at 60 miles per hour. If the total time was 4 hours, how far is it from here to Bristol?

38. A river has a current of 3 kilometers per hour. Find the speed of Jena's boat in still water if she travels 60 kilometers downstream in the same time as 36 kilometers upstream.

39. Leo can get to the lake using either the old road at 40 miles per hour, or the new road at 60 miles per hour. If both roads are the same length and he gets there 1 hour sooner on the new road, how far is it to the lake?

40. Olivia can ride her bike 4 miles per hour faster than Ted. If Olivia can go 30 miles in the same time that Ted can go 15 miles, what are their speeds?

Objective 5 **Solve applications about work rates.**

Solve each problem.

41. Rose and Mose want to paint a chair. Mose can do it in 3 hours while Rose can do it in 4 hours. How long will it take them working together?

42. Rod can do a job in 3 hours, while Karen requires 9 hours. How long will it take them if they work together?

43. Amy and Anne (identical twins) are cleaning up their room. Amy can do it in 8 hours, but Anne needs 11 hours. How long will it take them if they work together?

44. Jason can erase the blackboard in 11 minutes while Rebecca can do it in 13 minutes. How long will it take them if they work together?

45. A perfume factory has a vat for perfume. An inlet pipe can fill it in 10 hours, while an outlet pipe can empty it in 12 hours. How long will it take to fill the vat if both the outlet and inlet pipes are left open?

46. A vat of chocolate can be filled by an inlet pipe in 12 hours. An outlet pipe can empty the vat in 15 hours. How long will it take to fill the vat if both pipes are left open?

47. An inlet pipe can fill a barrel with wine in 8 hours, while an outlet pipe can empty it in 6 hours. Through an error both pipes are left open. How long will it take to empty the barrel?

48. Glenn can weed the garden in 3 hours while Mike can weed it in 5 hours. How long will it take them if they work together?

49. Michelle can type a report in 4 hours but Beth can do it in 1 hour. How long will it take them if they work together?

50. Sara can mow the lawn in 5 hours and Casey can do it in 6 hours. How long will it take if they work together?

8.5 Mixed Exercises

Find the value of the variable indicated.

51. If $F = \frac{9}{5}C + 32$, and $F = 98.6$, find C.

52. If $A = \frac{1}{2}h(B+b)$, $B = 18$, $h = 4$, and $A = 60$, find b.

53. If $\frac{1}{x} = \frac{1}{y} + \frac{1}{z}$, $x = 5$, and $y = 3$, find z.

54. If $\frac{mn}{p} = \frac{qr}{s}$, $m = 4$, $n = 2.5$, $q = 7.5$, $r = 2.5$, and $s = 1.5$, find p.

Solve each formula for the specified variable.

55. $C = \frac{5}{9}(F - 32)$ for F

57. $\frac{1}{p} + \frac{1}{8} = \frac{7}{r}$ for r

59. $F = f\left[\dfrac{v + v_0}{v - v_s}\right]$ for v_s

56. $\dfrac{6}{5x} - \dfrac{7}{8y} = \dfrac{3}{z}$ for x

58. $h = \dfrac{2A}{B + b}$ for B

60. $F = f\left[\dfrac{v + v_0}{v - v_s}\right]$ for v_0

Solve each problem.

61. Vince can clean the office he shares with Miklos in 5 hours, while Miklos can do the job in 4 hours. How long will it take them if they work together?

62. Mark's boat goes 10 miles per hour. Find the speed of the current of the river if he can go 30 miles upstream in the same time that he takes to go 70 miles downstream.

63. In basketball, Wendi hopes to complete 4 baskets for every 7 baskets she attempts. If she attempts 35 baskets, how many should she complete to fulfill her hopes?

64. If a vat of tomato juice at a cannery can be filled by an inlet pipe in 12 hours, and emptied by an outlet pipe in 18 hours, how long will it take to fill the vat if both pipes must remain open?

65. A stream has a current of 2 miles per hour. Find the speed of Barbara's boat in still water if it goes 16 miles downstream in the same time it takes to go 12 miles upstream.

66. Mike paid $5.12 sales tax on an item priced $64. Later he made a purchase that cost him a total of $51.84, including sales tax at the same rate. How much sales tax did he pay on the second item?

67. At 8:00 A.M. Richard left home to walk 5 miles to a friend's house. After visiting his friend for 1 hour, Richard's wife picked him up and drove home at a speed of 30 miles per hour. They arrived home at 10:25 A.M. What was Richard's speed walking to his friend's house?

68. Vivian has invested $4000 in an account that earned $180 in the past year. She reinvested the amount she earned in the same account without withdrawing any of the original amount. If the account earns income at the same rate during the next year, how much income will Vivian earn next year?

69. Lyle can fly her plane 250 miles against the wind in the same time that it takes her to fly 300 miles with the wind. If the wind blows at 25 miles per hour, find the speed of Lyle's plane in still air.

70. Suppose that Joe and Marty can clean their entire house in 6 hours, while their toddler, Midgie, just by being around, can completely mess it up in $1\frac{1}{2}$ hours. If Midgie comes home to a clean house after visiting her grandma and Joe and Marty start to clean up the minute she gets home, how long does it take until the house is in shambles?

Chapter 9

ROOTS, RADICALS, AND ROOT FUNCTIONS

9.1   Radical Expressions and Graphs

Objective 1   Find roots of numbers.

Find each root that is a real number. Use a calculator as needed.

1. $\sqrt[3]{27}$

2. $\sqrt[3]{216}$

3. $\sqrt[3]{-8}$

4. $\sqrt[4]{16}$

5. $\sqrt[4]{-81}$

6. $\sqrt[5]{243}$

7. $\sqrt[3]{-125}$

8. $\sqrt[5]{32}$

9. $\sqrt[4]{625}$

10. $\sqrt[5]{-1}$

11. $\sqrt[3]{3375}$

12. $\sqrt[3]{-64}$

13. $\sqrt[6]{64}$

14. $\sqrt[6]{729}$

15. $\sqrt[4]{256}$

16. $\sqrt[3]{-343}$

17. $\sqrt[4]{6561}$

18. $\sqrt[3]{512}$

19. $\sqrt[3]{-1331}$

20. $\sqrt[4]{14,641}$

Objective 2   Find principal roots.

Find each root that is a real number.

21. $\sqrt{9}$

22. $\sqrt[3]{64}$

23. $\sqrt[3]{-27}$

24. $\sqrt[4]{1}$

25. $\sqrt[5]{-243}$

26. $\sqrt[6]{x^{18}}$

27. $\sqrt{x^6}$

28. $\sqrt[5]{y^{10}}$

29. $\sqrt[5]{-a^{15}}$

30. $\sqrt[5]{-32}$

31. $\sqrt[4]{r^8}$

32. $\sqrt[9]{-c^{18}}$

33. $\sqrt[5]{3125}$

34. $\sqrt[3]{c^{27}}$

35. $-\sqrt[5]{32}$

36. $\sqrt[3]{a^6}$

37. $-\sqrt[5]{-32}$

38. $-\sqrt[3]{8}$

39. $\sqrt[15]{-1}$

40. $\sqrt{49}$

41. $-\sqrt{144}$

42. $-\sqrt{\dfrac{49}{16}}$

43. $-\sqrt{\dfrac{121}{25}}$

44. $\sqrt{(-5)^2}$

45. $\sqrt{t^2}$

46. $\sqrt{s^{16}}$

47. $\sqrt{p^{20}}$

48. $\sqrt{a^{36}}$

113

Objective 3 **Graph functions defined by radical expressions.**

Graph each function and give its domain and its range.

49. $f(x) = \sqrt{x+1}$

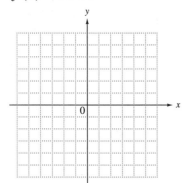

52. $f(x) = \sqrt{x} + 2$

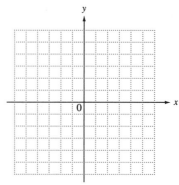

50. $f(x) = \sqrt{x-3}$

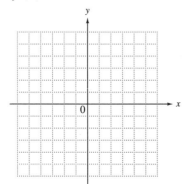

53. $f(x) = \sqrt[3]{x} - 2$

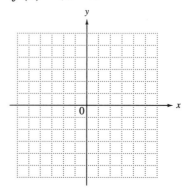

51. $f(x) = \sqrt{x} - 1$

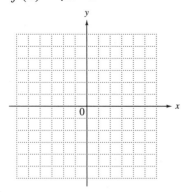

54. $f(x) = \sqrt[3]{x} + 2$

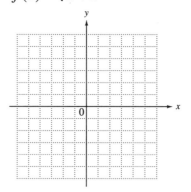

Objective 4 **Find nth roots of the nth powers.**

Simplify each root.

55. $\sqrt[4]{x^4}$

57. $\sqrt[5]{x^5}$

59. $\sqrt[6]{x^{12}}$

61. $-\sqrt[4]{x^{16}}$

56. $-\sqrt[4]{x^4}$

58. $-\sqrt[5]{x^5}$

60. $-\sqrt[5]{x^{10}}$

62. $\sqrt[7]{x^{21}}$

Objective 5   **Use a calculator to find roots.**

Use a calculator to find a decimal approximation for each radical. Round answers to three decimal places if necessary.

63. $-\sqrt{87}$

64. $\sqrt{76}$

65. $-\sqrt[3]{61}$

66. $-\sqrt[3]{35}$

67. $\sqrt[4]{42}$

68. $\sqrt[4]{84}$

69. $\sqrt{101}$

70. $-\sqrt{159}$

71. $\sqrt[5]{204}$

72. $\sqrt[3]{263}$

73. $-\sqrt{310}$

74. $\sqrt[6]{200}$

75. $\sqrt[3]{660}$

76. $\sqrt{870}$

77. $\sqrt{930}$

78. $-\sqrt[4]{130}$

79. $-\sqrt{370}$

80. $-\sqrt[5]{990}$

81. $\sqrt{123}$

82. $\sqrt[3]{28}$

9.1   Mixed Exercises

Find each root or use a calculator to find a decimal approximation. Round to three decimal places if necessary.

83. $\sqrt{196}$

84. $\sqrt[4]{4096}$

85. $\sqrt[3]{-3375}$

86. $\sqrt{384}$

87. $-\sqrt{333}$

88. $\sqrt[4]{720}$

89. $\sqrt{k^4}$

90. $\sqrt{d^2}$

91. $\sqrt[5]{-p^{10}}$

92. $\sqrt[3]{43}$

93. $-\sqrt{-39}$

94. $\sqrt[4]{45}$

9.2   Rational Exponents

In all exercises, assume that variables represent positive real numbers.

Objective 1   Use exponential notation for nth roots.

Evaluate each exponential.

1. $27^{1/3}$

2. $25^{1/2}$

3. $32^{1/5}$

4. $343^{1/3}$

5. $-81^{1/4}$

6. $(-81)^{1/4}$

7. $1000^{1/3}$

8. $(-8)^{1/3}$

9. $625^{1/4}$

10. $1024^{1/10}$

11. $-256^{1/4}$

12. $16^{1/4}$

13. $216^{1/3}$

14. $243^{1/5}$

15. $3375^{1/3}$

16. $512^{1/3}$

17. $81^{1/2}$

18. $125^{1/3}$

19. $1296^{1/4}$

20. $(-125)^{1/3}$

Objective 2   Define $a^{m/n}$.

Evaluate each exponential.

21. $27^{2/3}$

22. $25^{3/2}$

23. $81^{5/4}$

24. $27^{-2/3}$

25. $25^{-3/2}$

26. $243^{-2/5}$

27. $1000^{2/3}$

28. $-125^{-2/3}$

29. $16^{3/4}$

30. $216^{-2/3}$

31. $81^{-3/4}$

32. $729^{-1/6}$

33. $32^{3/5}$

34. $-8^{-2/3}$

35. $(-8)^{-2/3}$

36. $-625^{3/4}$

37. $512^{-2/3}$

38. $36^{5/2}$

39. $729^{5/6}$

40. $8^{-1/3}$

Objective 3   Convert between radicals and rational exponents.

Simplify each radical by rewriting it with a rational exponent. Write answers in radical form if necessary.

41. $\sqrt{x^{12}}$

42. $\sqrt[6]{4t^4}$

43. $\sqrt[3]{x^2} \cdot \sqrt[6]{x}$

44. $\sqrt{7^4}$

45. $\sqrt[8]{16x^{12}}$

46. $\sqrt[40]{y^{35}}$

47. $\sqrt[4]{5} \cdot \sqrt{5}$

48. $\sqrt[3]{2} \cdot \sqrt[4]{2}$

49. $\sqrt[3]{x} \cdot \sqrt{x}$

50. $\sqrt[5]{p^7}$

51. $\dfrac{\sqrt{r}}{\sqrt[3]{r}}$

52. $\sqrt[15]{27t^6}$

53. $\sqrt[3]{k^2} \cdot \sqrt[6]{k}$

54. $\sqrt[3]{p} \cdot \sqrt[5]{p^2}$

55. $\sqrt[3]{r^7}$

56. $\sqrt[4]{x^3} \cdot \sqrt[3]{x}$

57. $\sqrt[3]{x} \cdot \sqrt[4]{y} \cdot \sqrt{z}$

58. $\sqrt[4]{5} \cdot \sqrt{2}$

Objective 4   **Use the rules for exponents with rational exponents.**

Use the rules of exponents to simplify each expression. Write all answers with positive exponents.

59. $13^{4/5} \cdot 13^{6/5}$

60. $\dfrac{5^{3/7}}{5^{4/7}}$

61. $\dfrac{8^{3/4}}{8^{-1/4}}$

62. $(7^{2/3})^6$

63. $3^{1/2} \cdot 3^{3/2}$

64. $5^{3/4} \cdot 5^{9/4}$

65. $r^{2/3} \cdot r^{1/4}$

66. $y^{7/3} \cdot y^{-4/3}$

67. $\dfrac{a^{2/3} \cdot a^{-1/3}}{(a^{-1/6})^3}$

68. $(a^4)^{1/2} \cdot (a^6)^{1/3}$

69. $(5^{4/3})^6$

70. $(a^{-1})^{1/2}(a^{-3})^{-1/2}$

71. $x^{3/4} \cdot x^{5/6}$

72. $\dfrac{8^{3/5} \cdot 8^{-8/5}}{8^{-2}}$

73. $\dfrac{a^{4/5}}{a^{2/3}}$

74. $\left(\dfrac{c^6}{x^3}\right)^{2/3}$

75. $\dfrac{(x^{-3}y^2)^{2/3}}{(x^2 y^{-5})^{2/5}}$

76. $\dfrac{(x^{-1}y^{2/3})^3}{(x^{1/3}y^{1/2})^2}$

9.2   Mixed Exercises

Simplify. Write answers with positive exponents only.

77. $-243^{1/5}$

78. $128^{1/7}$

79. $\left(-\dfrac{32}{3125}\right)^{3/5}$

80. $\left(\dfrac{81}{256}\right)^{-3/4}$

81. $-1000^{2/3}$

82. $-32^{-3/5}$

83. $625^{-3/4}$

84. $100{,}000^{2/5}$

85. $\dfrac{w^{7/4}w^{-1/2}}{w^{5/4}}$

86. $\left(\dfrac{x^{1/4}}{x^{-3/4}}\right)^2$

Simplify by first converting to rational exponents. Write answers in radical form if necessary.

87. $\sqrt[3]{x^2} \cdot \sqrt[9]{x^4}$

88. $\dfrac{\sqrt[5]{x^2}}{\sqrt[3]{x}}$

9.3 Simplifying Radical Expressions

In all exercises, assume that variables represent positive real numbers.

Objective 1 **Use the product rule for radicals.**

Multiply.

1. $\sqrt{3} \cdot \sqrt{11}$

2. $\sqrt{14} \cdot \sqrt{5}$

3. $\sqrt{2} \cdot \sqrt{t}$

4. $\sqrt{7x} \cdot \sqrt{6t}$

5. $\sqrt{\dfrac{7}{c}} \cdot \sqrt{\dfrac{13}{w}}$

6. $\sqrt{\dfrac{11}{r}} \cdot \sqrt{\dfrac{3}{p}}$

7. $\sqrt[4]{7} \cdot \sqrt[4]{6}$

8. $\sqrt[6]{4t} \cdot \sqrt[6]{5t^4}$

9. $\sqrt[5]{6r^2t^3} \cdot \sqrt[5]{4r^2t}$

10. $\sqrt[3]{3} \cdot \sqrt[3]{7}$

11. $\sqrt{5} \cdot \sqrt{7}$

12. $\sqrt{11} \cdot \sqrt{15}$

13. $\sqrt{6} \cdot \sqrt{r}$

14. $\sqrt[3]{5} \cdot \sqrt[3]{x}$

15. $\sqrt[3]{7x} \cdot \sqrt[3]{5y}$

16. $\sqrt[4]{8} \cdot \sqrt[4]{15}$

17. $\sqrt[7]{8a^2t^3} \cdot \sqrt[7]{6at^3}$

18. $\sqrt[4]{2} \cdot \sqrt[4]{2x}$

Objective 2 **Use the quotient rule for radicals.**

Simplify each radical.

19. $\sqrt{\dfrac{25}{16}}$

20. $\sqrt{\dfrac{5}{36}}$

21. $\sqrt{\dfrac{x}{81}}$

22. $\sqrt{\dfrac{t^9}{25}}$

23. $\sqrt[3]{-\dfrac{27}{8}}$

24. $\sqrt[3]{\dfrac{45}{27}}$

25. $\sqrt{\dfrac{49}{100}}$

26. $\sqrt{\dfrac{6}{49}}$

27. $\sqrt[3]{\dfrac{125}{r^{15}}}$

28. $\sqrt[3]{-\dfrac{a^6}{125}}$

29. $\sqrt[3]{\dfrac{w^3}{216}}$

30. $\sqrt[3]{\dfrac{\ell^2}{27}}$

31. $\sqrt{\dfrac{r}{121}}$

32. $\sqrt[4]{\dfrac{p}{16}}$

33. $\sqrt[3]{-\dfrac{343}{125}}$

34. $\sqrt{\dfrac{z^4}{36}}$

35. $\sqrt[5]{\dfrac{7x}{32}}$

36. $\sqrt{\dfrac{15}{169}}$

Objective 3 **Simplify radicals.**

Express each radical in simplified form.

37. $\sqrt{27}$

38. $\sqrt{63}$

39. $\sqrt{75}$

40. $\sqrt{200}$

41. $\sqrt{135}$

42. $\sqrt[3]{24}$

43. $\sqrt[4]{32}$

44. $\sqrt{36t^5}$

45. $\sqrt{50r^5x}$

46. $\sqrt[3]{24y^6}$

47. $\sqrt[3]{54x^{11}}$

48. $\sqrt[3]{80x^9c^7}$

49. $\sqrt[3]{108a^5}$ **50.** $\sqrt{49r^9}$ **51.** $\sqrt[3]{270b^4c^8}$

Objective 4 **Simplify products and quotients of radicals with different indexes.**

Express each radical in simplified form.

52. $\sqrt[12]{x^{16}}$ **55.** $\sqrt[10]{x^{25}}$ **58.** $\sqrt[24]{z^{30}}$ **61.** $\sqrt[12]{9^3 x^6 y^9}$

53. $\sqrt[6]{x^9}$ **56.** $\sqrt[9]{9^3}$ **59.** $\sqrt[20]{x^{15}y^{10}}$ **62.** $\sqrt{6}\cdot\sqrt[3]{5}$

54. $\sqrt[24]{5^4}$ **57.** $\sqrt[42]{x^{28}}$ **60.** $\sqrt[6]{z^4 y^2}$ **63.** $\sqrt[3]{3}\cdot\sqrt[6]{7}$

Objective 5 **Use the Pythagorean formula.**

Find the missing length in each right triangle. Simplify the answer if necessary.

64.

65.

66.

67.

68.

69.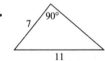

70.

71. 2 90° ∣— 10

72.

73.

Objective 6 **Use the distance formula.**

Find the distance between each pair of points.

74. $(2, 5)$ and $(-2, 3)$

75. $(3, 4)$ and $(-1, -2)$

76. $(1, -2)$ and $(4, -3)$

77. $(-1, 3)$ and $(5, -2)$

78. $(-1, -2)$ and $(-4, 3)$

79. $(4, 7)$ and $(-3, 8)$

80. $(6, -5)$ and $(9, 2)$

81. $(-4, 7)$ and $(5, -3)$

82. $\left(\sqrt{5}, -\sqrt{2}\right)$ and $\left(2\sqrt{5}, 3\sqrt{2}\right)$

83. $\left(4\sqrt{3}, 2\sqrt{5}\right)$ and $\left(3\sqrt{3}, -\sqrt{5}\right)$

84. $(2x, x + y)$ and $(x, x - y)$

85. $(a - b, a)$ and $(a + b, b)$

9.3   Mixed Exercises

Simplify.

86. $\sqrt{5a} \cdot \sqrt{6b}$

87. $\sqrt[4]{16x^{12}y^{10}}$

88. $\sqrt[5]{4w} \cdot \sqrt[5]{2w^3}$

89. $\sqrt[3]{54t^7}$

90. $\sqrt[3]{125}$

91. $\sqrt{3x^7y}$

92. $\sqrt[3]{\dfrac{27}{8}}$

93. $\sqrt{\dfrac{a^4}{625}}$

94. $\sqrt[3]{\dfrac{b^{13}q^6}{8}}$

95. $\sqrt{8x^3y^6z^{11}}$

96. $\sqrt[4]{t^3v^{12}}$

97. $\sqrt{3} \cdot \sqrt[5]{64}$

Find the missing length in the right triangle. Simplify the answer if necessary.

98.

99.

Find the distance between each pair of points.

100. $(-1, 6)$ and $(7, 4)$

101. $(5, -1)$ and $(2, -5)$

9.4 Adding and Subtracting Radical Expressions

Objective 1 Simplify radical expressions involving addition and subtraction.

Add or subtract. Assume that all variables represent positive real numbers.

1. $2\sqrt{7} + 3\sqrt{7}$

2. $3\sqrt{13} + 5\sqrt{52}$

3. $3\sqrt{48} + 5\sqrt{27}$

4. $2\sqrt{18} - 5\sqrt{32} + 7\sqrt{162}$

5. $5\sqrt[3]{24} + 4\sqrt[3]{81}$

6. $7\sqrt[4]{32} - 9\sqrt[4]{2}$

7. $5\sqrt{13} + 4\sqrt{13} - 6\sqrt{13}$

8. $3\sqrt{6} - 8\sqrt{6} - 5\sqrt{24}$

9. $\sqrt{48y} + \sqrt{12y} + \sqrt{27y}$

10. $\sqrt{98} - 2\sqrt{8} + \sqrt{32}$

11. $\sqrt[3]{81} + \sqrt[3]{24} + \sqrt[3]{192}$

12. $\sqrt[3]{625} + \sqrt[3]{135} - \sqrt[3]{40}$

13. $\sqrt{100x} - \sqrt{9x} + \sqrt{25x}$

14. $-2\sqrt[3]{81} + \sqrt[3]{24}$

Solve each problem. Give answers as simplified radical expressions.

15. Find the perimeter of a square with a side that measures $\sqrt{18}$ in. Find the area of the square.

16. Find the perimeter of a triangle with sides a, b, and c if $a = \sqrt{20}$ cm, $b = \sqrt{80}$ cm, and $c = \sqrt{125}$ cm.

17. Find the perimeter of a rectangle whose length is $\sqrt{12}$ ft and whose width is $\sqrt{27}$ ft. What is the area of the rectangle?

18. Use $A = \frac{1}{2}h(B+b)$ to find A if $h = 10\sqrt{5}$ cm, $B = 6\sqrt{3}$ cm, and $b = \sqrt{3}$ cm.

19. Use $A = \frac{1}{2}h(B+b)$ to find A if $b = 2\sqrt{5}$ m, $h = 3\sqrt{2}$ m, and $B = 4\sqrt{5}$ m.

9.4 Mixed Exercises

Add or subtract. Assume that all variables represent positive real numbers.

20. $5\sqrt{10} - \sqrt{10} - 3\sqrt{10}$

21. $3\sqrt{54} - 5\sqrt{24}$

22. $7\sqrt[3]{54} - 6\sqrt[3]{128}$

23. $6\sqrt[3]{135} + 3\sqrt[3]{40}$

24. $2\sqrt[3]{16r} + \sqrt[3]{54r} - \sqrt[3]{16r}$

25. $3\sqrt{8z} + 3\sqrt{2z} + \sqrt{32z}$

26. $3\sqrt{18z} + 2\sqrt{8z}$

27. $6\sqrt[3]{54z^3} + 5\sqrt[3]{16z^3}$

Solve each problem. Give answers as simplified radical expressions.

28. Find the width of a rectangle with length $3\sqrt{75}$ cm and perimeter $36\sqrt{3}$ cm. Find the area of the rectangle.

29. Use $A = \frac{1}{2}bh$ to find b if $A = 6\sqrt{105}$ in.^{2} and $h = 6\sqrt{3}$ in.

9.5   Multiplying and Dividing Radical Expressions

In all exercises, assume that variables represent positive real numbers.

Objective 1   Multiply radical expressions

Multiply each product, then simplify.

1. $\left(3+\sqrt{2}\right)\left(2+\sqrt{7}\right)$

2. $\left(\sqrt{10}+\sqrt{3}\right)\left(\sqrt{6}-\sqrt{11}\right)$

3. $\left(5+\sqrt{2}\right)\left(5-\sqrt{2}\right)$

4. $\left(\sqrt{5}+\sqrt{6}\right)\left(\sqrt{2}-4\right)$

5. $\left(\sqrt{2}+4\right)\left(\sqrt{2}-4\right)$

6. $\left(3\sqrt{3}+5\right)^{2}$

7. $\left(3\sqrt{2}-4\right)\left(2\sqrt{2}+7\right)$

8. $\left(\sqrt{2}-\sqrt{12}\right)^{2}$

9. $\left(\sqrt{5}+1\right)\left(\sqrt{5}-1\right)$

10. $\left(\sqrt{x}+y\right)\left(\sqrt{x}-y\right)$

11. $\left(2\sqrt{x}-3\right)\left(3\sqrt{x}-2\right)$

12. $\left(2+\sqrt[3]{5}\right)\left(2-\sqrt[3]{5}\right)$

Objective 2   Rationalize denominators with one radical term.

Rationalize the denominator in each expression.

13. $\dfrac{6}{\sqrt{5}}$

14. $\dfrac{15}{\sqrt{5}}$

15. $\dfrac{\sqrt{2}}{\sqrt{11}}$

16. $\dfrac{\sqrt{15}}{\sqrt{2}}$

17. $\dfrac{3}{\sqrt{18}}$

18. $\dfrac{3}{\sqrt{8}}$

19. $\dfrac{7}{\sqrt{75}}$

20. $-\dfrac{14}{\sqrt{27}}$

21. $\dfrac{12}{\sqrt{50}}$

22. $\dfrac{4}{\sqrt{6}}$

23. $\dfrac{7}{4\sqrt{3}}$

24. $\dfrac{\sqrt{3}}{8\sqrt{5}}$

25. $\dfrac{5}{2\sqrt{5}}$

26. $\dfrac{5}{\sqrt{10}}$

27. $\dfrac{3}{\sqrt{98}}$

Simplify.

28. $\sqrt{\dfrac{27}{48}}$

29. $\sqrt{\dfrac{36}{t}}$

30. $\sqrt{\dfrac{50}{r}}$

31. $\sqrt{\dfrac{162x^{4}}{t^{5}}}$

32. $\sqrt{\dfrac{8}{m}}$

33. $\sqrt{\dfrac{25}{y}}$

34. $\sqrt{\dfrac{5x}{8}}$

35. $\sqrt{\dfrac{4x^{2}}{3}}$

36. $\sqrt{\dfrac{18z}{7}}$

37. $\sqrt{\dfrac{4}{7y}}$

38. $\sqrt{\dfrac{5}{8x^2}}$

39. $\sqrt{\dfrac{12a^2}{5t^3}}$

40. $\sqrt{\dfrac{7y^2}{12b}}$

41. $\sqrt{\dfrac{19}{32}}$

42. $\sqrt{\dfrac{1}{5}}$

43. $\sqrt[3]{\dfrac{1}{9}}$

44. $\sqrt[3]{\dfrac{14}{243}}$

45. $\sqrt[3]{\dfrac{8}{100}}$

46. $\sqrt[3]{\dfrac{x}{y}}$

47. $\sqrt[3]{\dfrac{3}{8}}$

48. $\sqrt[3]{\dfrac{5x}{2}}$

49. $\sqrt[3]{\dfrac{7}{36}}$

50. $\sqrt[3]{\dfrac{4}{5}}$

51. $\sqrt[3]{\dfrac{c^3}{d^2}}$

52. $\sqrt[3]{\dfrac{t^6}{x^7}}$

53. $\sqrt[3]{\dfrac{r}{98}}$

54. $\sqrt[3]{\dfrac{m}{2x}}$

[Objective 3] **Rationalize denominators with binomials involving radicals.**

Rationalize each denominator.

55. $\dfrac{5}{7-\sqrt{3}}$

56. $\dfrac{-6}{\sqrt{7}+3}$

57. $\dfrac{5}{3-\sqrt{5}}$

58. $\dfrac{12}{6+\sqrt{8}}$

59. $\dfrac{18}{\sqrt{11}+\sqrt{2}}$

60. $-\dfrac{4}{\sqrt{7}-\sqrt{5}}$

61. $\dfrac{5}{\sqrt{3}+\sqrt{11}}$

62. $\dfrac{1}{4+\sqrt{5}}$

63. $\dfrac{1}{4-\sqrt{7}}$

64. $\dfrac{2}{\sqrt{2}-\sqrt{3}}$

65. $\dfrac{7}{\sqrt{3}-1}$

66. $-\dfrac{6}{\sqrt{6}-\sqrt{3}}$

67. $-\dfrac{5}{\sqrt{5}-\sqrt{3}}$

68. $\dfrac{\sqrt{3}}{\sqrt{5}-\sqrt{2}}$

69. $\dfrac{\sqrt{6}}{\sqrt{13}+\sqrt{5}}$

[Objective 4] **Write radical quotients in lowest terms.**

Write each quotient in lowest terms.

70. $\dfrac{12-3\sqrt{2}}{3}$

71. $\dfrac{30-45\sqrt{3}}{20}$

72. $\dfrac{9+6\sqrt{15}}{12}$

73. $\dfrac{35-7\sqrt{6}}{7}$

74. $\dfrac{7-\sqrt{98}}{14}$

75. $\dfrac{5+2\sqrt{75}}{25}$

76. $\dfrac{12+18\sqrt{3}}{6}$

78. $\dfrac{8-2\sqrt{12}}{4}$

80. $\dfrac{6+2\sqrt{5}}{2}$

77. $\dfrac{2x-\sqrt{8x^2}}{4x}$

79. $\dfrac{50+\sqrt{80x}}{10}$

81. $\dfrac{16-12\sqrt{72}}{24}$

9.5   Mixed Exercises

Multiply, then simplify each product.

82. $\sqrt{3}\left(\sqrt{18}-2\sqrt{12}\right)$

83. $\left(\sqrt[3]{2}+1\right)\left(\sqrt[3]{2}-1\right)$

Rationalize the denominator in each expression.

84. $-\dfrac{15}{\sqrt{8}}$

86. $\dfrac{3\sqrt{2}}{\sqrt{t}}$

88. $\dfrac{3\sqrt{x}}{2\sqrt{t}+\sqrt{x}}$

85. $\dfrac{12}{\sqrt{27}}$

87. $\dfrac{3-\sqrt{7}}{\sqrt{3}-7}$

89. $\dfrac{\sqrt{t}}{\sqrt{t}+3}$

Simplify.

90. $\sqrt{\dfrac{27}{t^3}}$

93. $\sqrt{\dfrac{5a}{3b}}$

96. $\sqrt[3]{\dfrac{s^2}{6}}$

91. $\sqrt{\dfrac{3}{10}}$

94. $\sqrt{\dfrac{x^2}{y^3}}$

97. $\sqrt[3]{\dfrac{t^{15}}{x}}$

92. $\sqrt{\dfrac{c}{x^2}}$

95. $\sqrt[3]{\dfrac{a^4}{b}}$

98. $\sqrt[3]{\dfrac{2x}{3z}}$

Write each quotient in lowest terms.

99. $\dfrac{12-3\sqrt{8}}{9}$

101. $\dfrac{12+\sqrt{20}}{10}$

103. $\dfrac{3+\sqrt{18x}}{3}$

100. $\dfrac{7y-\sqrt{98y^5}}{14y}$

102. $\dfrac{5+10\sqrt{2}}{5}$

104. $\dfrac{5\sqrt{5}-\sqrt{25}}{35}$

9.6   Solving Equations with Radicals

Objective 1   Solve radical equations using the power rule.

Solve each equation.

1. $\sqrt{t} = 5$

2. $\sqrt{4x+1} = 3$

3. $\sqrt{3t+7} = 5$

4. $\sqrt{9c+9} = 9$

5. $\sqrt{7x-6} = 8$

6. $\sqrt{6x+1} = 1$

7. $\sqrt{2m+6} = 6$

8. $\sqrt{r+8} = 3$

9. $\sqrt{2p+5} = 5$

10. $\sqrt{4x-19} = 5$

11. $\sqrt{5p-5} = 5$

12. $\sqrt{x-7} = 3$

13. $\sqrt{2q-1} = 9$

14. $\sqrt{12h+1} = 7$

15. $\sqrt{6x+25} = 25$

16. $\sqrt{3w+4} = 7$

17. $\sqrt{9c+1} = 8$

18. $\sqrt{3x+1} = 8$

19. $\sqrt{x+2} = 3$

20. $\sqrt{12x+16} = 16$

21. $\sqrt{t} = -6$

22. $\sqrt{a} - 6 = -2$

23. $\sqrt{x-7} - 4 = 0$

24. $\sqrt{c+3} + 7 = 0$

25. $\sqrt{t+1} - 4 = 0$

26. $\sqrt{2r+6} - 4 = 0$

27. $\sqrt{12p+1} + 7 = 0$

28. $\sqrt{y-4} - 3 = 0$

29. $\sqrt{2x+5} - 3 = 0$

30. $\sqrt{5x-4} - 1 = 0$

31. $\sqrt{x+3} - 5 = 0$

32. $\sqrt{2x+3} - 5 = 0$

33. $\sqrt{7y-5} - 3 = 0$

34. $\sqrt{5r-4} - 9 = 0$

35. $\sqrt{m} + 7 = -1$

36. $\sqrt{x+13} - 5 = 2$

Objective 2   Solve radical equations that require additional steps.

Solve each equation.

37. $\sqrt{6a-23} = 3a - 11$

38. $\sqrt{2x+7} = 2x + 1$

39. $\sqrt{4x+17} = x + 3$

40. $\sqrt{x-3} = x - 3$

41. $\sqrt{x-1} = x - 7$

42. $\sqrt{13-3x} = x - 5$

43. $\sqrt{8x+33} = x + 3$

44. $\sqrt{41-16x} = x - 6$

45. $\sqrt{30-5x} = x - 6$

46. $\sqrt{7x+39} = x + 3$

47. $\sqrt{7y+15} = 2y + 3$

48. $\sqrt{3q-8} = q - 2$

49. $\sqrt{33-8r} = 2r - 3$

50. $\sqrt{44-20x} = -8x$

51. $\sqrt{25-8x} = x - 2$

52. $\sqrt{12x+4} = x + 2$

53. $\sqrt{58-11w} = w + 2$

54. $\sqrt{27-18v} = 2v - 3$

55. $\sqrt{t+2} - \sqrt{t-3} = 1$

56. $\sqrt{c+1} + \sqrt{4c-3} = 5$

57. $\sqrt{3k+7} + \sqrt{k+1} = 2$

58. $\sqrt{x+1} + \sqrt{7x+4} = 7$

59. $\sqrt{2p+5} + \sqrt{p+51} = 8$

60. $\sqrt{4w-3} + \sqrt{w+2} = 8$

61. $\sqrt{5y+4} - \sqrt{2y+2} = 1$

62. $\sqrt{x+4} + \sqrt{9-4x} = 5$

63. $\sqrt{x-4} + \sqrt{x+11} = 5$

64. $\sqrt{3x+3} - \sqrt{2x+3} = 1$

65. $\sqrt{5x+9} - \sqrt{3x+4} = 1$

66. $\sqrt{2-r} + \sqrt{r+11} = 5$

67. $\sqrt{5c+6} - \sqrt{c+3} = 3$

68. $\sqrt{x+12} + \sqrt{14-c} = 6$

69. $\sqrt{7x+4} - \sqrt{x+6} = 2$

70. $\sqrt{11x+22} - \sqrt{9x+19} = 1$

71. $\sqrt{3p+10} + \sqrt{p+2} = 6$

72. $\sqrt{k+10} + \sqrt{2k+19} = 2$

Objective 3 Solve radical equations with indexes greater than 2.

Solve each equation.

73. $\sqrt[3]{7x-5} - \sqrt[3]{3x+7} = 0$

74. $\sqrt[4]{3x-3} = 3$

75. $\sqrt[3]{t^2+5t+15} = \sqrt[3]{t^2}$

76. $\sqrt[4]{3k+2} + 2 = 0$

77. $\sqrt[5]{2t+1} - 1 = 0$

78. $\sqrt[4]{8x+5} = \sqrt[4]{7x+7}$

79. $\sqrt[3]{5r-6} - \sqrt[3]{3r+4} = 0$

80. $\sqrt[5]{2w+5} = \sqrt[5]{7w}$

81. $\sqrt[3]{4x-4} = \sqrt[3]{3x+13}$

82. $\sqrt[4]{c^2+2c+18} = \sqrt[4]{c^2}$

83. $\sqrt[5]{b-1} - 2 = 0$

84. $\sqrt[3]{5x-12} = \sqrt[3]{6x-15}$

85. $\sqrt[5]{5a+1} - \sqrt[5]{2a-11} = 0$

86. $\sqrt[3]{6x-4} = \sqrt[3]{2x+8}$

87. $\sqrt[3]{2a-63} + 5 = 0$

88. $\sqrt[3]{3x-11} = \sqrt[3]{2x+10}$

89. $\sqrt[3]{8-x} - 2 = 0$

90. $\sqrt[3]{x^3-125} + 5 = 0$

9.6 Mixed Exercises

Solve each equation.

91. $\sqrt{3q-1} + 5 = 0$

92. $\sqrt{x} - 8 = 0$

93. $\sqrt{28x-6} = 2x+3$

94. $\sqrt[3]{4t-1} - 3 = 0$

95. $\sqrt{2w+25} + \sqrt{2w+16} = 9$

96. $\sqrt{7r+8} - \sqrt{r+1} = 5$

97. $\sqrt[3]{r^2+3r+15} - \sqrt[3]{r^2} = 0$

98. $\sqrt{h+7} = 2$

99. $\sqrt{30t+19} = 2t+5$

100. $\sqrt{x+12} - 4 = 0$

9.7   Complex Numbers

Objective 1 Simplify numbers of the form $\sqrt{-b}$, where $b > 0$.

Write each number as a product of a real number and i. Simplify all radical expressions.

1. $\sqrt{-49}$ **6.** $\sqrt{-50}$ **11.** $\sqrt{-60}$ **16.** $\sqrt{-1080}$

2. $\sqrt{-36}$ **7.** $-\sqrt{-63}$ **12.** $\sqrt{-450}$ **17.** $-\sqrt{-625}$

3. $-\sqrt{-100}$ **8.** $\sqrt{-120}$ **13.** $-\sqrt{-125}$ **18.** $-\sqrt{-162}$

4. $\sqrt{-6}$ **9.** $\sqrt{-18}$ **14.** $-\sqrt{-72}$

5. $\sqrt{-22}$ **10.** $-\sqrt{-27}$ **15.** $\sqrt{-99}$

Multiply or divide as indicated.

19. $\sqrt{-6} \cdot \sqrt{-6}$

20. $\sqrt{-3} \cdot \sqrt{-15}$

27. $\dfrac{\sqrt{-125}}{\sqrt{-5}}$

32. $\dfrac{\sqrt{-80}}{\sqrt{5}}$

21. $\sqrt{-21} \cdot \sqrt{-7}$

28. $\dfrac{\sqrt{-16}}{\sqrt{2}}$

33. $\dfrac{\sqrt{-200}}{\sqrt{-8}}$

22. $\sqrt{-14} \cdot \sqrt{3}$

29. $\dfrac{\sqrt{-28}}{\sqrt{7}}$

34. $\dfrac{\sqrt{-42} \cdot \sqrt{-6}}{\sqrt{-7}}$

23. $\sqrt{-3} \cdot \sqrt{7}$

24. $\sqrt{2} \cdot \sqrt{-5} \cdot \sqrt{3}$

30. $\dfrac{\sqrt{-18}}{\sqrt{-2}}$

35. $\dfrac{\sqrt{-56} \cdot \sqrt{-6}}{\sqrt{16}}$

25. $\sqrt{-6} \cdot \sqrt{-3} \cdot \sqrt{2}$

26. $\sqrt{-5} \cdot \sqrt{-3} \cdot \sqrt{-7}$

31. $\dfrac{\sqrt{-45}}{\sqrt{-5}}$

36. $\dfrac{\sqrt{-10} \cdot \sqrt{7}}{\sqrt{25}}$

Objective 2 Recognize imaginary complex numbers.

*The real numbers are a subset of the complex numbers. Classify each of the following complex numbers as **real** or **imaginary**.*

37. $7 - 3i$ **40.** $\sqrt{3} - i\sqrt{5}$ **42.** $-\dfrac{2}{3}i$ **44.** $\sqrt{3} + \sqrt{5}$

38. $-i\sqrt{10}$ **45.** $5 - i\sqrt{2}$

39. $\sqrt{5}$ **41.** $-\dfrac{1}{3}$ **43.** $\dfrac{3}{4} + \dfrac{5}{4}i$ **46.** $i\sqrt{7}$

Objective 3 Add and subtract complex numbers.

Add or subtract as indicated. Write your answers in standard form.

47. $(5 + 7i) + (-2 + 4i)$ **48.** $(-2 + 9i) + (10 - 3i)$

49. $(5-2i)+(9+0i)$

50. $4+(3+6i)$

51. $(8+3i)-(5+3i)$

52. $(-2+3i)-(5+i)$

53. $(-7-2i)-(-3-3i)$

54. $7-(2-3i)$

55. $(7-3i)+(2-6i)$

56. $4i-(9+5i)+(2+3i)$

57. $(3+3i)-(8+4i)$

58. $(-1-5i)+(2+5i)$

59. $(7-9i)-(5-6i)$

60. $(5+3i)+(5-3i)$

61. $\left(\sqrt{3}-2i\sqrt{2}\right)+\left(2\sqrt{3}-2i\sqrt{2}\right)$

62. $[(8+4i)-(5-3i)]+(4-2i)$

Objective 4 **Multiply complex numbers.**

Multiply.

63. $(2+5i)(3-i)$

64. $(2-3i)(2+7i)$

65. $(4-6i)(2+6i)$

66. $(5-3i)(5+3i)$

67. $(7+2i)(7-2i)$

68. $(3-2i)(2-3i)$

69. $(1+3i)^2$

70. $\left(\sqrt{2}-i\sqrt{3}\right)^2$

71. $(8+2i)(3-5i)$

72. $(2+i)(3-i)$

Objective 5 **Divide complex numbers.**

Write each quotient in the form a + bi.

73. $\dfrac{1+i}{2-i}$

74. $\dfrac{5+2i}{9-4i}$

75. $\dfrac{7+9i}{4-i}$

76. $\dfrac{3-i}{i}$

77. $\dfrac{4+i}{3-2i}$

78. $\dfrac{3-2i}{2+i}$

79. $\dfrac{6-i}{2-3i}$

80. $\dfrac{3i}{2-i}$

81. $\dfrac{2-i}{4i}$

82. $\dfrac{1+i}{(2-i)(2+i)}$

Objective 6 **Find powers of *i*.**

Find each power of i.

83. i^{11}

84. i^{17}

85. i^{48}

86. i^{42}

87. i^{-13}

88. i^{-7}

89. i^{100}

90. i^{62}

91. i^{307}

92. i^{236}

93. i^{-100} **94.** i^{75} **95.** i^{-23} **96.** i^{115} **97.** i^{-21}

9.7 Mixed Exercises

Simplify.

98. $\sqrt{-125}$

99. $\sqrt{-4}$

100. i^9

101 i^{-3}

Perform the indicated operations. Write answers in the form a + bi.

102. $\sqrt{-7} \cdot \sqrt{-35}$

103. $\sqrt{-8} \cdot \sqrt{-8}$

104. $(8-5i)+(3+i)$

105. $(12+2i)-(-1+i)$

106. $(12+2i)(-1+i)$

107. $5+(-3+4i)$

108. $(3+2i)(5-i)$

109. $(2-5i)(2+5i)$

110. $\dfrac{6+i}{2-3i}$

111. $\dfrac{5-9i}{4-3i}$

Chapter 10

QUADRATIC EQUATIONS, INEQUALITIES, AND FUNCTIONS

10.1 The Square Root Property and Completing the Square

Objective 1 Learn the square root property.

Use the square root property to solve each equation.

1. $p^2 = 81$

2. $x^2 - 1 = 8$

3. $x^2 + 9 = 16$

4. $4 - y^2 = 0$

5. $x^2 = 13$

6. $11 - s^2 = 9$

7. $8 = 2q^2$

8. $3x^2 + 11 = 38$

9. $12 - 4y^2 = 8$

10. $r^2 - 2 = 8$

Objective 2 Solve quadratic equations of the form $(ax+b)^2 = c$ by using the square root property.

Use the square root property to solve each equation.

11. $(a+1)^2 = 64$

12. $(r+4)^2 = 3$

13. $(c-5)^2 = 36$

14. $(2x+3)^2 = 5$

15. $(3x+7)^2 = 25$

16. $(7c-1)^2 - 5 = 1$

17. $(a+6)^2 = 36$

18. $(4x+6)^2 = 1$

19. $(r-1)^2 = 16$

20. $(s-3)^2 = 36$

21. $(3x-2)^2 = 25$

22. $(2x-1)^2 - 16 = 0$

Objective 3 Solve quadratic equations by completing the square.

Solve each equation by completing the square.

23. $x^2 + 4x + 3 = 0$

24. $2x^2 - 5x = 0$

25. $2x^2 - 4x = 1$

26. $m(m + 7) + 12 = 0$

27. $p^2 + p - 11 = 0$

28. $3t^2 + 4t - 1 = 0$

29. $a^2 - 2a - 2 = 0$

30. $z^2 - \dfrac{z}{2} - \dfrac{7}{4} = 0$

31. $(2r+1)^2 + 2(2r+1) - 3 = 0$

32. $x^2 - 12x + 27 = 0$

33. $(5m+2)^2 - 9(5m+2) - 36 = 0$

34. $x^2 + 3x + 2 = 0$

35. $x^2 - 9x + 8 = 0$

36. $c^2 - 4c - 9 = 0$

Objective 4   Solve quadratic equations with imaginary solutions.

Find the imaginary solutions of each equation.

37. $x^2 + 1 = 0$

38. $9 + y^2 = 0$

39. $p^2 = -16$

40. $9 = -(6q - 7)^2$

41. $k^2 + 25 = 0$

42. $(m + 1)^2 = -36$

43. $4x^2 + 25 = 0$

44. $121 + q^2 = 0$

45. $-81 - k^2 = 0$

46. $49 + 16(2w + 4)^2 = 0$

10.1 Mixed Exercises

Solve each equation.

47. $(a + 5)^2 = 7$

48. $b^2 - 225 = 0$

49. $z^2 = 27$

50. $(2s + 4)^2 = 7$

51. $(x - 3)^2 = -1$

52. $(4a + 5)^2 = -12$

53. $\dfrac{9x^2}{2} = 50$

54. $2a^2 = 20$

55. $x^2 + 3x = 0$

56. $x^2 + 7x + 1 = 0$

57. $2a^2 + 7a - 13 = 0$

58. $x^2 - 2x + 3 = 0$

59. $r^2 - 5r + 4 = 0$

60. $x^2 - 3 = \dfrac{1}{2}x$

10.2 The Quadratic Formula

Objective 1 Derive the quadratic formula.

Objective 2 Solve quadratic equations using the quadratic formula.

Use the quadratic formula to solve each equation.

1. $x^2 - 7x + 6 = 0$

2. $x^2 - 12x + 27 = 0$

3. $x^2 + 5x - 14 = 0$

4. $p^2 - 3p - 40 = 0$

5. $r^2 - 8r + 16 = 0$

6. $5t^2 - 13t + 6 = 0$

7. $6m^2 - 17m + 12 = 0$

8. $3w^2 + 6w - 24 = 0$

9. $16x^2 - 9 = 0$

10. $(z + 2)^2 = 2(5z - 2)$

11. $5m^2 + 5m - 1 = 0$

12. $4p(p + 1) = 1$

13. $3x^2 + 1 = 6x$

14. $2x^2 = 4x - 3$

15. $\dfrac{m}{2m - 1} = \dfrac{2m + 3}{15}$

16. $x^2 + 1 = \dfrac{13x}{6}$

Objective 2 Use the discriminant to determine the number and type of solutions.

Use the discriminant to determine whether the solutions for each equation are

A. *two rational numbers,*
C. *two irrational numbers,*

B. *one rational number,*
D. *two imaginary numbers.*

Do not actually solve.

17. $5a^2 - 4a + 1 = 0$

18. $2k^2 + 2k + 3 = 0$

19. $p^2 - 2p + 4 = 0$

20. $t^2 + 5t + 4 = 0$

21. $2y^2 + 4y + 8 = 0$

22. $3r^2 + 5r + 1 = 0$

23. $m^2 - 4m + 4 = 0$

24. $5y^2 - 5y + 2 = 0$

25. $4t^2 + 12t + 9 = 0$

26. $z^2 + 6z + 3 = 0$

27. $2r^2 = 4r + 3$

28. $3p^2 = 2p + 5$

29. $2m^2 - 4m = 8$

30. $n^2 + n = 2$

10.2 Mixed Exercises

Use the quadratic formula to solve each equation.

31. $t^2 - 2t + 3 = 0$

32. $x^2 - x - 3 = 0$

33. $r^2 = 12r + 28$

34. $4x^2 = -4x + 8$

Use the discriminant to determine whether the solutions for each equation are

A. *two rational numbers,*   B. *one rational number,*
C. *two irrational numbers,*   D. *two imaginary numbers.*

Do not actually solve.

35. $12r^2 - 40r + 25$ **37.** $5p^2 - 7p - 12$ **39.** $16x^2 - 40x + 25$

36. $6x^2 - 3x - 10$ **38.** $18p^2 + 46p - 24$ **40.** $16x^2 - 12x + 9$

10.3 Equations Quadratic in Form

Objective 1 Solve an equation with fractions by writing it in quadratic form.

Solve each equation. Check your solutions.

1. $1 + \dfrac{1}{x} = \dfrac{6}{x^2}$

2. $\dfrac{x}{2x+15} = \dfrac{1}{3x-2}$

3. $5 + \dfrac{6}{m+1} = \dfrac{14}{m}$

4. $\dfrac{7}{x^2} + 6 = -\dfrac{23}{x}$

5. $2 + \dfrac{9}{x} = \dfrac{2}{x+1}$

6. $4 - \dfrac{8}{x-1} = -\dfrac{35}{x}$

7. $2 = \dfrac{1}{x} + \dfrac{28}{x^2}$

8. $9 - \dfrac{12}{x} = -\dfrac{4}{x^2}$

9. $1 - \dfrac{2}{x} - \dfrac{15}{x^2} = 0$

10. $\dfrac{5x}{x+1} + \dfrac{6}{x+2} = \dfrac{3}{(x+1)(x+2)}$

11. $\dfrac{3}{x} + \dfrac{1}{x-3} = \dfrac{7}{4}$

12. $\dfrac{5}{x} + \dfrac{1}{2x+7} = -\dfrac{2}{3}$

13. $\dfrac{2m}{m-5} + \dfrac{7}{m+1} = 0$

14. $2 + \dfrac{3}{p} = \dfrac{5}{p^2}$

15. $x = \dfrac{1}{x-3} + \dfrac{17}{3}$

16. $\dfrac{2}{x^2} + \dfrac{1}{x} = 1$

Objective 2 Use quadratic equations to solve applied problems.

Solve each problem.

17. The perimeter of a rectangle is 24 inches. The area is 32 square inches. Find the length and the width of the rectangle.

18. Amy rows her boat upstream 6 miles and back in $2\frac{6}{7}$ hours. The speed of the current is 2 miles per hour. How fast can she row?

19. Bill can row 3 miles per hour in still water. It takes him 3 hours and 36 minutes to go 3 miles upstream and return. Find the speed of the current.

20. Two pipes together can fill a large tank in 6 hours. One of the pipes, used alone, takes 5 hours longer than the other to fill the tank. How long would each pipe used alone take to fill the tank?

21. The distance from Appletown to Medina is 45 miles, as is the distance from Medina to Westmont. Karl drove from Westmont to Medina, stopped at Medina for a hamburger, and then drove on to Appletown at 10 miles per hour faster. Driving time for the entire trip was 99 minutes. Find Karl's speed from Westmont to Medina.

22. Two pipes together can fill a large tank in 10 hours. One of the pipes, used alone, takes 15 hours longer than the other to fill the tank. How long would each pipe used alone take to fill the tank?

23. A jet plane traveling at a constant speed goes 1200 miles with the wind, then turns around and travels for 1000 miles against the wind. If the speed of the wind is 50 miles per hour and the total flight takes 4 hours, find the speed of the plane.

24. The sum of the reciprocal of a number and the reciprocal of 5 more than the number is $\frac{11}{24}$. What is the number?

25. A man rode a bicycle for 12 miles and then hiked an additional 8 miles. The total time for the trip was 5 hours. If his rate when he was riding the bicycle was 10 miles per hour faster than his rate walking, what was each rate?

26. The sum of a number and its reciprocal is 5.2. What is the number?

Objective 3   **Solve an equation with radicals by writing it in quadratic form.**

Solve each equation. Check your solutions.

27. $x = \sqrt{x+2}$

28. $\sqrt{7y-10} = y$

29. $\sqrt{2y} = \sqrt{6-y}$

30. $k - \sqrt{8k-15} = 0$

31. $x = \sqrt{\dfrac{x+3}{2}}$

32. $y = \sqrt{\dfrac{1-2y}{8}}$

33. $\sqrt{3}y = \sqrt{28y-49}$

34. $p = \sqrt{\dfrac{7p-1}{12}}$

35. $\sqrt{\dfrac{25r-1}{144}} = r$

36. $y = \sqrt{y+12}$

Objective 4   **Solve an equation that is quadratic in form by substitution.**

Solve each equation. Check your solutions.

37. $x^4 - 25x^2 + 144 = 0$

38. $16m^4 = 25m^2 - 9$

39. $(x+1)^2 = 10(x+1) + 75$

40. $c^4 - 13c^2 + 36 = 0$

41. $(m+5)^2 + 6(m+5) + 8 = 0$

42. $(3-r)^2 = -3(3-r) + 18$

43. $\sqrt{x} = x - 6$

44. $m^{-2} - m^{-1} - 12 = 0$

45. $x^4 = -x^2 + 20$

46. $a^4 + 63 = 16a^2$

47. $t^4 = \dfrac{21t^2 - 5}{4}$

48. $\dfrac{1}{(x+6)^2} - \dfrac{7}{2(x+6)} = -\dfrac{3}{2}$

49. $x^4 - 20x^2 + 36 = 0$

50. $m^4 - 16m^2 = 80$

51. $x^4 - 11x^2 + 24 = 0$

52. $x^4 - 12x^2 + 20 = 0$

53. $m^4 - 10m^2 + 24 = 0$

54. $p^4 - 12p^2 + 27 = 0$

55. $x^4 - 2x^2 + 1 = 0$

56. $x^4 = 3x^2$

58. $36x^4 - 3601x^2 + 100 = 0$

57. $\dfrac{1}{9}(r^2 + 3)^2 = r^2 + 1$

10.3 Mixed Exercises

Solve each problem. Round each answer to the nearest tenth.

59. Andrew can do a job in 3 hours less time than Emily. If they can finish the job together in 10 hours, how long will it take Andrew working alone?

60. Two pipes together can fill a large tank in 15 hours. One of the pipes, used alone, takes 12 hours longer than the other to fill the tank. How long would each pipe used alone take to fill the tank?

61. Cara rowed her boat across and back Lake Bend in 3 hours. If her rate returning was 2 miles per hour less than the rate going, and if the distance each way was 7 miles, find her rate going.

Solve each equation. Check your solutions.

62. $10 - \dfrac{7}{c^2} = -\dfrac{33}{c}$

63. $\dfrac{x-2}{x} + \dfrac{1}{x-1} = \dfrac{5}{6}$

64. $1 + \dfrac{49}{2x} = \dfrac{15}{x+1}$

65. $\dfrac{1}{p} - \dfrac{6}{p^2} = -1$

66. $(m+1)^2 - \dfrac{8}{3}(m+1) = 1$

67. $(t^2 - 3t)^2 = 14(t^2 - 3t) - 40$

68. $(\sqrt{x} + 3)^2 = 8(\sqrt{x} + 3) - 12$

69. $4m^4 + 12 = 19m^2$

10.4 Formulas and Further Applications

Objective 1   **Solve formulas for variables involving squares and square roots.**

Solve each equation for the indicated variable. (Leave ± in your answers.)

1. $D = \sqrt{kh}$ for k

2. $F = \dfrac{kl}{\sqrt{d}}$ for d

3. $p = \sqrt{\dfrac{kl}{g}}$ for k

4. $p = \dfrac{yz}{\sqrt{6}}$ for z

5. $s = 30\sqrt{\dfrac{a}{p}}$ for p

6. $a = \sqrt{bc} + 1$ for c

7. $y = \dfrac{1}{2}gt^2$ for t

8. $F = \dfrac{mx}{t^2}$ for t

9. $F = \dfrac{1}{2}kx^2$ for x

Objective 2   **Solve applied problems using the Pythagorean formula.**

Solve each problem.

10. A 13-foot ladder is leaning against a building. The distance from the bottom of the ladder to the building is 2 feet more than twice the distance from the top of the ladder to the ground. How far is the bottom of the ladder from the building?

11. A ladder is leaning against a building so that the top is 8 feet above the ground. The length of the ladder is 2 feet less than twice the distance of the bottom of the ladder from the building. Find the length of the ladder.

12. Two cars left an intersection at the same time, one heading south, the other heading east. Some time later the car traveling south had gone 18 miles farther than the car headed east. At that time they were 90 miles apart. How far had each car traveled?

13. Two cars left an intersection at the same time, one heading north, the other heading west. Later they were exactly 95 miles apart. The car headed west had gone 38 miles less than twice as far as the car headed north. How far had each car traveled?

14. A child flying a kite has let out 45 feet of string to the kite. The distance from the kite to the ground is 9 feet more than the distance from the child to a point directly below the kite. How high up is the kite?

15. The height of a kite above the ground is 4 feet less than twice the distance from the person flying the kite to a point directly below it. The length of the string to the kite is 68 feet. How high is the kite?

16. The longest side of a right triangle is 4 centimeters longer than the next longest side. The third side is 16 centimeters in length. Find the length of the longest side.

17. The longest side of a right triangle is 2 feet more than the middle side, and the middle side is 1 foot less than twice the shortest side. Find the length of the shortest side.

18. The two shorter sides of a right triangle have lengths that differ by 4 meters. The longest side is 20 meters. Find the shortest side.

19. The longest side of a right triangle is 1 foot more than the middle side. The shortest side is 7 feet. Find the longest side.

20. The width of a rectangle is 14 centimeters. The diagonal is 2 centimeters more than the length. Find the length of the rectangle.

21. The diagonal of a rectangle is 25 inches, and the length is 3 inches more than three times the width. What is the length of the rectangle?

Objective 3 **Solve applied problems using area formulas.**

Solve each problem.

22. A rectangle has a length 1 meter less than twice its width. If 1 meter is cut from the length and added to the width, the figure becomes a square with an area of 16 square meters. Find the dimensions of the original rectangle.

23. The length of a rectangle is 4 inches more than its width. If 2 inches is taken from the length and added to the width, the figure becomes a square with an area of 196 square inches. What are the dimensions of the original figure?

24. The area of a square is 81 square centimeters. If the same amount is added to one dimension and removed from the other, the resulting rectangle has an area 9 centimeters less than the area of the square. How much is added and subtracted?

25. A square has an area of 81 square inches. If the same amount is added to one side and subtracted from an adjacent side, the resulting rectangle has an area of 77 square inches. Find the dimensions of the rectangle.

26. A rectangular piece of cardboard has a length that is 3 inches longer than the width. A square 1.5 inches on a side is cut from each corner. The sides are then turned up to form an open box with a volume of 162 cubic inches. Find the dimensions of the original piece of cardboard.

27. A piece of plastic in the shape of a rectangle has a length 10 inches less than twice the width. A square 4 inches on a side is cut out of each corner and the sides turned up to form an open box with a volume of 160 cubic inches. Find the dimensions of the finished box.

28. An open box is to be made from a rectangular piece of tin by cutting 2-inch squares out of the corners and folding up the sides. The length of the finished box is to be twice the width. The volume of the box will be 100 cubic inches. Find the dimensions of the rectangular piece of tin.

29. To make an open box, 3-centimeter squares are cut from the corners of a rectangular piece of cardboard and the sides folded up. The length of the cardboard is 6 centimeters more than its width. The volume of the finished box will be 120 cubic centimeters. Find the dimensions of the piece of cardboard.

30. The floor of a room 14 feet by 18 feet is to be tiled with a border of even width around the edges. How wide should the border be if the region inside the border is to have an area of 140 square feet?

31. A rectangular garden has an area of 12 feet by 5 feet. A gravel path of equal width is to be built around the garden. How wide can the path be if there is enough gravel for 138 square feet?

32. A picture 9 inches by 12 inches is to be mounted on a piece of mat board so that there is an even width of mat all around the picture. How wide will the matted border be if the area of the mounted picture is 238 square inches?

33. A rug is to fit in a room so that a border of even width is left on all four sides. If the room is 16 feet by 20 feet and the area of the rug is 165 square feet, how wide to the nearest tenth of a foot will the border be?

Objective 4 **Solve applied problems using quadratic functions as models.**

Solve each problem. Round answers to the nearest tenth.

34. A population of microorganisms grows according to the function $p(x) = 100 + .2x + .5x^2$, where x is given in hours. How many hours does it take to reach a population of 250 microorganisms?

35. The charge potential for a certain experiment can be modeled by the function $c(x) = 5 - .1x - .2x^2$, where x is given in minutes. When does the potential equal 4.5?

36. The position of an object moving in a straight line is given by $s(t) = t^2 - 8t$, where s is in feet and t is in seconds. How long will it take the object to move 10 feet?

37. An object is thrown downward from a tower 280 feet high. The distance the object has fallen at time t in seconds is given by $s(t) = 16t^2 + 68t$. How long will it take the object to fall 100 feet?

10.4 Mixed Exercises

Solve each equation for the indicated variable. (Leave ± in your answers.)

38. $pq = t^2 - pt$ for t

40. $p^2q^2 + pkq = k^2$ for q

39. $xy = m^2 + xm$ for m

41. $b^2a^2 + 2bca = c^2$ for a

Solve each problem.

42. A doghouse 2 feet by 4 feet is to be built with a cement path around it of equal width on all sides. The area available for the doghouse and path is 120 square feet. How wide will the path be?

43. The length of a rectangle is 5 centimeters less than the diagonal, and the width is 5 centimeters less than the length. Find the width of the rectangle.

44. The length of a rectangle is 2 inches more than twice the width. The diagonal is 1 inch more than the length. Find the diagonal.

45. A rug is to fit in a room so that a border of even width is left on all four sides. If a room is 12 feet by 15 feet and the area of the rug is 108 square feet, how wide will the border be?

46. The profit from the sale of x items is given by the function $P(x) = 2x^2 - 10x - 100$. What is the minimum number of items that must be sold for the profit to exceed $1000?

10.5 Graphs of Quadratic Functions

Objective 1 Graph a quadratic function.

Objective 2 Graph parabolas with horizontal and vertical shifts.

Identify the vertex and graph each parabola.

1. $f(x) = x^2 - 2$

2. $f(x) = x^2 + 2$

3. $f(x) = x^2 + 3$

4. $f(x) = x^2 - 4$

5. $f(x) = 2 - x^2$

6. $f(x) = 5 - x^2$

7. $f(x) = (x + 2)^2$

8. $f(x) = (x - 3)^2$

9. $f(x) = (x + 3)^2 - 1$

10. $f(x) = (x - 2)^2 + 1$

Objective 3 Predict the shape and direction of a parabola from the coefficient of x^2.

For each quadratic function, tell whether the graph opens up or down and whether the graph is wider, narrower, or the same shape as the graph of $f(x) = x^2$.

11. $f(x) = \dfrac{1}{2}x^2$

12. $f(x) = 3x^2$

13. $f(x) = -2x^2$

14. $f(x) = -\dfrac{4}{3}x^2 - 1$

15. $f(x) = \dfrac{3}{5}x^2 + 5$

16. $f(x) = \dfrac{1}{3}(x - 2)^2$

17. $f(x) = -2(x + 1)^2$

18. $f(x) = -\dfrac{1}{3}(x + 3)^2 - 4$

19. $f(x) = \dfrac{5}{4}(x - 1)^2 + 7$

20. $f(x) = 4 - x^2$

Objective 4 Find a quadratic function to model data.

Tell whether a linear or quadratic function would be a more appropriate model for each set of graphed data. If linear, tell whether the slope should be positive or negative. If quadratic, tell whether the coefficient a of x^2 should be positive or negative.

21.
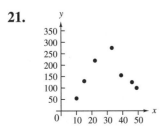

22.
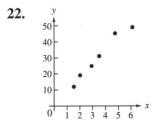

23.
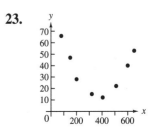

10.5 Mixed Exercises

Identify the vertex and graph each parabola.

24. $f(x) = (x-1)^2$

26. $f(x) = (x-3)^2 - 1$

25. $f(x) = (x+2)^2 + 3$

27. $f(x) = (x+3)^2$

For each quadratic function, tell whether the graph opens up or down and whether the graph is wider, narrower, or the same shape as the graph of $f(x) = x^2$. Find each vertex.

28. $f(x) = 4x^2$

29. $f(x) = -x^2 + 2$

30. $f(x) = \frac{1}{3}(x-1)^2$

31. $f(x) = \frac{3}{2}(x+1)^2 - 2$

32. $f(x) = -\frac{1}{2}(x+3)^2$

33. $f(x) = 4(x-3)^2 + 1$

10.6 More about Parabolas; Applications

Objective 1 Find the vertex of a vertical parabola.

Objective 2 Graph a quadratic function.

Graph each function and find its vertex.

1. $f(x) = x^2 + 6x + 10$

2. $f(x) = x^2 - 6x + 4$

3. $f(x) = -x^2 + 8x - 10$

4. $f(x) = x^2 - 3x + 2$

5. $f(x) = 3x^2 + 6x + 2$

6. $f(x) = -2x^2 + 4x + 1$

7. $f(x) = \frac{1}{2}x^2 + 2x + 3$

8. $f(x) = \frac{5}{4}x^2 + 5x + 3$

Objective 3 Use the discriminant to find the number of x-intercepts of a vertical parabola.

Use the discriminant to determine the number of x-intercepts of the graph of each function.

9. $f(x) = 6x^2 - 3x + 1$

10. $f(x) = 3x^2 + 3x + 2$

11. $f(x) = x^2 - 2x + 4$

12. $f(x) = x^2 + 5x + 4$

13. $f(x) = 2x^2 + 4x + 8$

14. $f(x) = 3x^2 + 5x + 1$

15. $f(x) = x^2 - 4x - 4$

16. $f(x) = 5x^2 - 5x + 2$

17. $f(x) = 4x^2 + 12x + 9$

18. $f(x) = x^2 + 5x + 2$

Objective 4 Use quadratic functions to solve problems involving maximum or minimum value.

Solve each problem.

19. A businessman has found that his daily profits are given by $P(x) = -2x^2 + 100x + 2400$, where x is the number of units sold each day. Find the number he should sell daily to maximize his profit. What is the maximum profit?

20. The same businessman has daily costs of $C(x) = x^2 - 50x + 1625$, where x is the number of units sold each day. How many units must be sold to minimize his cost? What is the minimum cost?

21. Jean sells ceramic pots. She has weekly costs of $C(x) = x^2 - 100x + 2700$, where x is the number of pots she sells each week. How many pots should she sell to minimize her costs? What is the minimum cost?

A projectile is fired upward so that its distance (in feet) above the ground t seconds after firing is as given. Find the maximum height it reaches and the number of seconds it takes to reach that height.

22. $s(t) = -16t^2 + 64t$

24. $s(t) = -16t^2 + 48t + 250$

23. $s(t) = -16t^2 + 32t$

25. $s(t) = -16t^2 + 80t + 156$

Solve each problem.

26. The length and width of a rectangle have a sum of 48. What width will produce the maximum area?

27. The length and width of a rectangle have a sum of 64. What width will produce the maximum area?

28. The perimeter of a rectangle is 24. What length will produce the maximum area?

Objective 5 **Graph horizontal parabolas.**

Graph each parabola. Give the domain and range.

29. $x = 2y^2$

33. $x = y^2 + 4y + 4$

30. $x = -2y^2$

34. $x = -y^2 + 4y - 4$

31. $x = -y^2 + 2$

35. $x = y^2 - 4y + 7$

32. $x = y^2 - 3$

36. $3x = y^2 - 6y + 6$

10.6 Mixed Exercises

Graph each parabola and find its vertex.

37. $y = -\frac{1}{3}x^2 - 2x - 4$

38. $x = y^2 + 6y + 5$

39. $x = -y^2 - 6y - 10$

40. $y = 2x^2 + 4x - \frac{1}{2}$

Use the discriminant to determine the number of x-intercepts of the graph of each function.

41. $f(x) = -x^2 + 5x - 3$

42. $f(x) = 2x^2 - 3x + 2$

Solve each problem.

43. Victor Retzlaff has found that the profits (in dollars) of his video store are approximately given by $p(x) = -x^2 + 16x + 34$, where x is the number of units of videos that he should rent daily to produce the maximum profit. How many units of videos should he rent daily to produce the maximum profit? Find the maximum profit.

44. The perimeter of a rectangle is 16. What length will produce the maximum area?

45. Of all pairs of numbers whose sum is 92, find the pair with the maximum product.

10.7 Quadratic and Rational Inequalities

Objective 1 Solve quadratic inequalities.

Solve each inequality, and graph the solution set.

1. $(x-2)(x+3) \geq 0$

2. $(y-2)(y+3) < 0$

3. $(m-5)(m+2) < 0$

4. $(r+3)(r-2) \geq 0$

5. $k^2 + 7k + 12 > 0$

6. $a^2 - a - 2 \leq 0$

7. $2y^2 < y + 3$

8. $2m^2 - 5m > 12$

9. $6r^2 + 7r + 2 > 0$

10. $8k^2 + 10k > 3$

11. $(x-1)^2 \geq -3$

12. $(x-3)^2 < 0$

13. $(2k+5)^2 \leq -1$

14. $(3p+2)^2 \geq -6$

15. $(2a+9)^2 < 0$

16. $(3x-2)^2 < -1$

Objective 2 Solve polynomial inequalities of degree 3 or more.

Solve each inequality, and graph the solution set.

17. $(x+1)(x-2)(x+4) \leq 0$

18. $(y+2)(y-1)(y-2) < 0$

19. $(k+5)(k-1)(k+3) \leq 0$

20. $(x-1)(x-3)(x+2) \geq 0$

21. $(y-4)(y+3)(y+1) \leq 0$

22. $(p-6)(p-4)(p-2) > 0$

23. $(2x-1)(2x+3)(3x+1) \leq 0$

24. $(4b+1)(6b-1)(3b-7) > 0$

25. $(4q-3)(2q-7)(3q-10) \geq 0$

26. $(z+1)(z-1)(3z-7) < 0$

Objective 3 Solve rational inequalities.

Solve each inequality, and graph the solution set.

27. $\dfrac{7}{x-1} \leq 1$

28. $\dfrac{2}{p+3} \leq 1$

29. $\dfrac{5}{x+1} \geq 1$

30. $\dfrac{4}{m+2} > 3$

31. $\dfrac{-2}{p+1} > 3$

32. $\dfrac{-1}{z-3} \leq 2$

33. $\dfrac{5}{y+2} < 0$

34. $\dfrac{3}{a+4} > 0$

35. $\dfrac{6}{3r-2} \geq 1$

36. $\dfrac{4}{3q+5} \geq -3$

37. $\dfrac{-5}{2x-3} \leq 2$

38. $\dfrac{-3}{4m-3} \geq 1$

39. $\dfrac{y}{y+1} \geq 3$

41. $\dfrac{2p-1}{3p+1} \leq 1$

43. $\dfrac{5}{x-3} \leq -1$

40. $\dfrac{r}{r-2} \geq 4$

42. $\dfrac{z+2}{z-3} \leq 2$

44. $\dfrac{4z}{3z-5} < -3$

10.7 Mixed Exercises

Solve each inequality, and graph the solution set.

45. $15k^2 + 2 \leq 11k$

46. $6z^2 \geq 11z + 10$

47. $4x^2 - 9 \leq 0$

48. $16r^2 - 9 \geq 0$

49. $8p^2 + 2p > 1$

50. $6m^2 + 17m < 3$

51. $(m-2)(m+1)(m+3) \geq 0$

52. $(y+2)(y-2)(y-3) < 0$

53. $(2r-1)(r-3)(r+1) < 0$

54. $(4a-3)(a-1)(a-3) \geq 0$

55. $(3k-1)(2k+3)(k-2) > 0$

56. $(5m-4)(2m+5)(m+1) \leq 0$

57. $\dfrac{m-3}{m} \geq 3$

58. $\dfrac{r+2}{r} \leq 5$

59. $\dfrac{2p-3}{3p} \leq -1$

60. $\dfrac{4k-3}{2k} \geq -2$

61. $(y+7)^2 + 6 \leq 0$

62. $(2x-1)^2 + 3 \geq 0$

Chapter 11

EXPONENTIAL AND LOGARITHMIC FUNCTIONS

11.1 Inverse Functions

Objective 1 Decide whether a function is one-to-one and, if it is, find its inverse.

If the function is one-to-one, find its inverse.

1. $\{(1, 0), (2, 0), (3, 5), (4, 1)\}$

2. $\{(-1, 1), (1, -1), (2, 1)\}$

3. $\{(2, -1), (-2, 1), (1, 3), (-1, -3)\}$

4. $\{(6, -3), (4, -2), (2, -1), (0, 0)\}$

5. $\{(4, 0), (2, 3), (0, 0), (3, 5)\}$

6. $\{(1, 5), (2, 5), (3, 5)\}$

7. $\{(-1, 1), (-2, 2), (-3, 3)\}$

8. $\{(-3, 1), (-2, 2), (-1, 3), (0, 4)\}$

9. $\{(3, 2), (-3, -2), (2, 3), (-2, -3)\}$

10. $\{(5, -1), (4, 7), (6, 3), (3, 3)\}$

Objective 2 Use the horizontal line test to determine whether a function is one-to-one.

Use the horizontal line test to determine whether each function is one-to-one.

11.

14.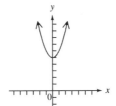

17.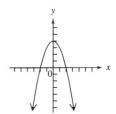

12.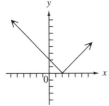

15.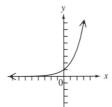

18.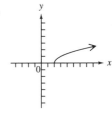

13.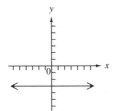

16.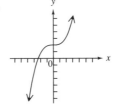

Objective 3 Find the equation of the inverse of a function.

If the function is one-to-one, find its inverse.

19. $f(x) = 2x - 5$

20. $f(x) = 3x - 5$

21. $f(x) = x^2 - 1$

22. $f(x) = 1 - 2x^2$

23. $f(x) = \sqrt{x-1}, \; x \geq 1$

24. $f(x) = 2\sqrt{3x}, \; x \geq 0$

25. $f(x) = x^3 - 1$

27. $f(x) = \dfrac{x^2 + 3}{2}$

28. $f(x) = \dfrac{3}{x - 1}$

26. $f(x) = 2x^3 - 3$

Objective 4 **Graph f^{-1} from the graph of f.**

If the function is one-to-one, graph the function f and its inverse f^{-1} on the same set of axes.

29.

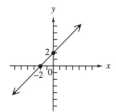

32.

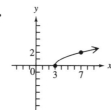

35.

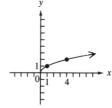

30.

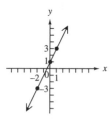

33.

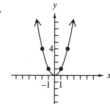

36.

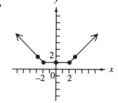

31.

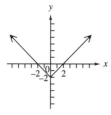

34.

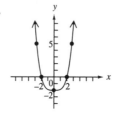

11.1 Mixed Exercises

If the function is one-to-one, find its inverse.

37. $\{(3, 5), (2, 9), (4, 7)\}$

41. $f(x) = 4 - 2x$

38. $\{(0, 0), (1, 1), (-1, -1), (2, 2), (-2, -2)\}$

42. $f(x) = \sqrt{x + 2},\ x \geq -2$

39. $\{(2, 4), (-1, 1), (0, 0), (1, 1), (2, 4)\}$

43. $f(x) = 2x^2 + 3$

40. $\{(-3, -1), (-2, 0), (-1, 1), (0, 2)\}$

44. $f(x) = x^3 - 5$

Use the horizontal line test to determine whether each function is one-to-one. If it is, graph the function f and its inverse f^{-1} on the same set of axes.

45.

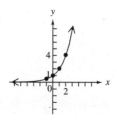

46.

11.2 Exponential Functions

Objective 1 Define exponential functions.

Decide whether or not each function defines an exponential function.

1. $f(x) = 2^x$ **4.** $f(x) = 3^x$ **7.** $f(x) = 2^{x+1}$ **10.** $f(x) = 1^x$

2. $f(x) = x^2$ **5.** $f(x) = (-3)^x$ **8.** $f(x) = 3^{2x}$

3. $f(x) = x + 2$ **6.** $f(x) = (x+1)^3$ **9.** $f(x) = 2x^3$

Objective 2 Graph exponential functions.

Graph each exponential function.

11. $f(x) = 3^x$

14. $f(x) = \left(\dfrac{1}{8}\right)^x$

12. $f(x) = -4^x$

15. $f(x) = 2^{1-x}$

13. $f(x) = 2^{-x}$

16. $f(x) = -2^{x-2}$

Objective 3 Solve exponential equations of the form $a^x = a^k$ for x.

Solve each equation.

17. $16^x = 64$ **20.** $4^{2x} = 8$ **23.** $4^{k+2} = 32$ **26.** $100^{2+t} = 1000$

18. $27^k = 9$ **21.** $9^y = 3$ **24.** $25^{2x-1} = 5$

19. $25^p = 625$ **22.** $10^{2x} = 100$ **25.** $25^{1-t} = 5$

Objective 4 Use exponential functions in applications involving growth or decay.

Solve each problem.

27. The population of Canadian geese that spend the summer at Gemini Lake each year has been growing according to the function

$$f(x) = 56(2)^{.2x},$$

where x is the time in years from 1978. Find the number of geese in 1994.

28. The diameter in inches of a tree during a certain period grew according to the function

$$f(x) = 2.5(9)^{.05x},$$

where x was the number of years after the start of this growth period. Find the diameter of the tree after 10 years.

29. A culture of a certain kind of bacteria grows according to the function

$$f(x) = 3650(2)^{.8x},$$

where x is the time in hours after 12 noon. Find the number of bacteria in the culture at 12 noon.

30. An industrial city in Pennsylvania has found that its population is declining according to the function
$$f(x) = 70,000(2)^{-.01x},$$
where x is the time in years from 1900. What is the city's anticipated population in the year 2000?

31. A sample of a radioactive substance with mass in grams decays according to the function
$$f(x) = 100(10)^{-.2x},$$
where x is the time in hours after the original measurement. Find the mass of the substance after 10 hours.

32. When a bactericide is placed in a certain culture of bacteria, the number of bacteria decreases according to the function
$$f(x) = 3200(4)^{-.1x},$$
where x is the time in hours. Find the number of bacteria in the culture after 20 hours.

33. Suppose the number of bacteria present in a certain culture after t minutes is given by the function
$$Q(t) = 500(2)^{.5t}.$$
Find the number of bacteria present after 2 minutes.

34. The population of Evergreen Park is now 16,000. The population t years from now is given by the function
$$P(t) = 16,000(2)^{t/10}.$$
What will the population be 40 years from now?

11.2 Mixed Exercises

Solve each equation.

35. $4^x = 8$

36. $25^{-2x} = 3125$

37. $16^{-x+1} = 8$

38. $5^{-x} = \dfrac{1}{5}$

39. $\left(\dfrac{1}{3}\right)^x = 27$

40. $\left(\dfrac{3}{4}\right)^x = \dfrac{16}{9}$

Graph each exponential function..

41. $f(x) = 4^{2x-3}$

42. $f(x) = -3^{-x}$

Solve each problem.

43. Corinna's savings grows according to the function
$$A(t) = P(1.01)^{4t},$$
where P is the amount of her original deposit and t is the number of years since the deposit was made. If $P = \$10,000$, how much will she have in 25 years?

44. The production of an oil well, in barrels, is decreasing according to the function
$$f(t) = 1,000,000(2)^{-.4t},$$
where t is the number of years after the well was drilled. Find the production after 5 years.

11.3 Logarithmic Functions

Objective 1 Define a logarithm.

Simplify. (Example: $\log_3 9 = 2.$)

1. $\log_2 8$

2. $\log_8 64$

3. $\log_3 \dfrac{1}{3}$

4. $\log_3 \sqrt{3}$

5. $\log_{1/2} 4$

6. $\log_{10} .0001$

7. $\log_7 \sqrt{7}$

8. $\log_4 2$

9. $\log_5 \dfrac{1}{25}$

10. $\log_{81} 27$

Objective 2 Convert between exponential and logarithmic forms.

Complete the chart.

	Exponential Form	Logarithmic Form
11.	$3^2 = 9$	
12.	$5^{1/3} = \sqrt[3]{5}$	
13.		$\log_4 \dfrac{1}{16} = -2$
14.		$\log_{16} 2 = \dfrac{1}{4}$
15.	$10^{-2} = \dfrac{1}{100}$	
16.		$\log_5 25 = 2$
17.		$\log_9 3 = \dfrac{1}{2}$
18.	$9^{1/2} = 3$	
19.		$\log_{10} .001 = -3$
20.	$2^{-7} = \dfrac{1}{128}$	

Objective 3 Solve logarithmic equations of the form $\log_a b = k$ for a, b, or k.

Solve each equation.

21. $\log_2 64 = p$ **24.** $\log_k 27 = 3$ **27.** $\log_n .01 = -2$ **30.** $\log_{1/4} \dfrac{1}{4} = p$

22. $\log_5 x = -1$ **25.** $\log_2 t = -5$ **28.** $\log_{1/2} r = -2$

23. $\log_m 25 = 2$ **26.** $\log_4 16 = y$ **29.** $\log_4 x = 0$

Objective 4 **Define and graph logarithmic functions.**

Graph each logarithmic function.

31. $y = \log_2 x$ **34.** $y = \log_{1/2} x$ **37.** $y = \log_{10} x$

32. $y = \log_9 x$ **35.** $y = \log_{1/9} x$ **38.** $y = \log_2(-x)$

33. $y = -\log_4 x$ **36.** $y = -\log_{1/4} x$

Objective 5 **Use logarithmic functions in applications of growth or decay.**

Solve each problem.

39. After black squirrels were introduced to Williams Park, their population grew according to the function

$$f(x) = 10\log_5(x+20),$$

where x is the number of months after the squirrels were introduced. Find the number of squirrels after 5 months.

40. A manufacturer receives revenue in dollars for selling x units of an item according to the function

$$f(x) = 200\log_3(x+1).$$

Find the revenue for selling 26 units.

41. Under certain conditions, the velocity v of the wind in centimeters per second is given by

$$v = 300\log_2\left(\frac{10x}{7}\right),$$

where x is the height in centimeters above the ground. Find the wind velocity at 11.2 centimeters above the ground.

42. A decibel is a measure of the loudness of a sound. A very faint sound is assigned an intensity of I_0, then another sound is given an intensity I found in terms of I_0, the faint sound. The decibel rating of the sound is given in decibels by

$$d = 10\log_{10}\frac{I}{I_0}.$$

Find the decibel rating of rock music that has intensity $I = 100,000,000,000I_0$.

43. The number of students not completing intermediate algebra is given by the function
$$f(s) = 4\log_7(8s+9),$$
where s is the number of sections of the class that is offered. If there are 5 sections of intermediate algebra this semester, how many students will not complete the course?

44. The number of fish in an aquarium is given by the function
$$f(t) = 10\log_2(3t+2),$$
where t is time in months. Find the difference in the number of fish present between $t=0$ and $t=10$.

45. After black squirrels were introduced to Sherman Park, their population grew according to the function
$$f(x) = 12\log_5(2x+5),$$
where x is the number of months after the squirrels were introduced. How many more squirrels were there in the park 10 months after being introduced than there were originally?

46. A company analyst has found that the number of applicants for new mortgages after a major advertising blitz is given by the function
$$A(x) = 50\log_2(2x+2),$$
where x is time in weeks after the blitz was started. Find the number of applicants for $x=7$.

47. Sales (in thousands) of a new product are approximated by
$$S = 125 + 20\log_2(30t+4) + 30\log_4(35t-6),$$
where t is the number of years after the product is introduced. Find the total sales 2 years after the product is introduced.

48. A population of mites in a laboratory is growing according to the function
$$p = 50\log_3(20t+7) - 25\log_9(80t+1),$$
where t is the number of days after a study is begun. Find the number of mites present 1 day after the beginning of the study.

11.3 Mixed Exercises

49. Write $\left(\frac{1}{2}\right)^{-3} = 8$ in logarithmic form.

50. Write $\log_5 .0016 = -4$ in exponential form.

Solve each equation.

51. $x = \log_{32} 8$

52. $\log_{1/3} r = -4$

53. $\log_5 1 = x$

54. $\log_a 4 = \frac{1}{2}$

55. Graph the function $y = \log_3 3x$.

Solve each problem.

56. The population of foxes in an area t months after the foxes were introduced there is approximated by the function

$$F(t) = 500\log_{10}(2t+3).$$

Find the number of foxes in the area when the foxes were first introduced into the area.

57. The number of fish in an aquarium is given by the function

$$f(t) = 8\log_5(2t+5),$$

where t is time in months. Find the number of fish present when $t = 10$.

58. A company analyst has found that total sales in thousands of dollars after a major advertising campaign are given by

$$S(x) = 100\log_2(x+2),$$

where x is time in weeks after the campaign was introduced. Find the sales when the campaign was introduced.

11.4 Properties of Logarithms

Objective 1 Use the product rule for logarithms.

Use the product rule to express each logarithm as a sum of logarithms, or as a single number if possible.

1. $\log_3 (6)(5)$

2. $\log_2 (5)(3)$

3. $\log_7 5m$

4. $\log_2 6xy$

5. $\log_6 6r$

6. $\log_3 2p$

Use the product rule to express each sum as a single logarithm.

7. $\log_4 7 + \log_4 3$

8. $\log 4 + \log 3$

9. $\log_7 11y + \log_7 2y + \log_7 3y$

10. $\log_7 8r^2 + \log_7 5r^2 + \log_7 3r$

Objective 2 Use the quotient rule for logarithms.

Use the quotient rule for logarithms to express each logarithm as a difference of logarithms, or as a single number if possible.

11. $\log_2 \dfrac{7}{9}$

12. $\log_4 \dfrac{5}{8}$

13. $\log_3 \dfrac{m}{n}$

14. $\log \dfrac{p}{r}$

15. $\log_6 \dfrac{k}{3}$

16. $\log_3 \dfrac{10}{x}$

17. $\log_2 \dfrac{8}{m}$

18. $\log_5 \dfrac{5}{x}$

Use the quotient rule for logarithms to express each difference as a single logarithm.

19. $\log_2 7q^4 - \log_2 5q^2$

20. $\log 9x^3 - \log 3x^2$

21. $\log_7 60r^3 - \log_7 100r^7$

22. $\log_9 40y^5 - \log_9 20y^7$

Objective 3 Use the power rule for logarithms.

Use the power rule for logarithms to rewrite each logarithm.

23. $\log_5 3^2$

24. $\log_3 4^3$

25. $\log_2 5^3$

26. $\log_m 2^7$

27. $\log_b \sqrt{5}$

28. $\log_3 \sqrt[3]{7}$

29. $\log_2 \sqrt{2}$

30. $\log_4 \sqrt[3]{4}$

31. $\log_2 \sqrt[3]{8}$

32. $\log_5 125^{1/3}$

Objective 4   **Use properties to write alternative forms of logarithmic expressions.**

Use the properties of logarithms to express each logarithm as a sum or difference of logarithms, or as a single number if possible.

33. $\log_2 4p^3$

34. $\log_3 9x^3$

35. $\log_a \sqrt[3]{2k}$

36. $\log_b \dfrac{2r}{r-1}$

37. $\log_2 \dfrac{4}{3}$

38. $\log_3 \dfrac{5}{9}$

39. $\log_5 \dfrac{7m^3}{8y}$

40. $\log_7 \dfrac{8r^7}{3a^3}$

Use the properties of logarithms to express each sum or difference of logarithms as a single logarithm, or as a single number if possible.

41. $\log 2x + \log 7x$

42. $\log_a 2r + \log_a 4r^2$

43. $\log_b 3pq - \log_b 2p^2$

44. $\log 4k^2 j - \log 3kj^2$

45. $\log_4 10y + \log_4 3y - \log_4 6y^3$

46. $\log_6 14m + \log_6 7m^2 - \log_6 14m^4$

47. $\log_2 (x-1) + \log_2 (x+1) - \log_2 (x^2-1)$

48. $\log_6 (r+4) + \log_6 (r-3) - \log_6 (r^2+r-12)$

11.4 Mixed Exercises

Use the properties of logarithms to express each logarithm as a sum or difference of logarithms, or as a single number if possible.

49. $\log_2 8p$

50. $\log_3 \sqrt{27}$

51. $\log_4 \dfrac{4}{9}$

52. $\log_5 k^4$

53. $\log_2 32^{2/5}$

54. $\log_5 \sqrt{3p}$

55. $\log_7 \dfrac{3}{7}$

56. $\log_4 \dfrac{3m}{m+2}$

Use the properties of logarithms to express each sum of difference of logarithms as a single logarithm, or as a single number if possible.

57. $\log_2 6y^3 - \log_2 3y^3$

58. $\log_3 2q + \log_3 5q^3$

59. $\log_5 8y + \log_5 2y$

60. $\log_{10} 1000p^5 - \log_{10} 100p^5$

61. $2\log_5 m + 3\log_5 m^2 - 4\log_5 m^3$

62. $2\log_2 y^2 + \log_2 y - 2\log_2 y^3$

11.5 Common and Natural Logarithms

Objective 1 Evaluate common logarithms using a calculator.

Use a calculator to find each logarithm. Give an approximation to four decimal places.

1. log 57.23

2. log 8

3. log 843.71

4. log .091419

5. log 280,037

6. log 798.886

7. log .00003184

8. log 61.000958

9. log .000958

10. log 87,123

11. log 22

12. log .3501

13. log 767

14. log 5489.62

15. log .000829

16. log .001

17. log 1,031,057

18. log 4.0014

Objective 2 Use common logarithms in applications.

Find the pH of solutions with the given hydronium ion concentrations. Round answers to the nearest tenth.

19. 2.8×10^{-6}

20. 5.6×10^{-8}

21. 2.1×10^{-7}

22. 1.7×10^{-9}

23. 6.2×10^{-5}

24. 7.4×10^{-11}

Find the hydronium ion concentration of solutions with the given pH values.

25. 2.9

26. 3.4

27. 5.2

28. 1.3

29. 6.5

30. 10.2

Objective 3 Evaluate natural logarithms using a calculator.

Find each natural logarithm. Give an approximation to four decimal places.

31. ln .12

32. ln 100

33. ln 6

34. ln 428

35. ln .013

36. ln 69

37. ln 4

38. ln .102

39. ln 874

40. ln 76.3

41. ln .01

42. ln .00214

Objective 4 Use natural logarithms in applications.

Solve each problem.

The population of a small town from 1975–1995 is approximated by the function

$$P(t) = 600e^{.01t},$$

where $t = 0$ represents 1975. Find the population in the given years.

43. 1975

44. 1978

45. 1985

46. 1995

A radioactive substance, in grams, is decaying so that the amount present at time t in days is given by the function

$$Q(t) = 100e^{-.03t}.$$

Find the amount present to the nearest tenth of a gram after the given number of days.

47. Initially   **48.** 10   **49.** 30   **50.** 60

The number of bacteria in a certain culture is approximated by

$$B(t) = 10,000e^{.05t},$$

where t is the time in hours. Find the population present to the nearest hundred after the given number of hours.

51. Initially   **52.** 3   **53.** 4   **54.** 24

11.5 Mixed Exercises

Use a calculator to find each logarithm. Give an approximation to four decimal places.

55. log .093621   **56.** ln 50   **57.** ln .000806   **58.** log 60,183.006

Solve each problem.

59. Find the pH of a substance with hydronium ion concentration of 3.9×10^{-9}. Round the answer to the nearest tenth.

60. Find the hydronium ion contraction of a substance with pH of 4.8. Round the answer to the nearest tenth.

61. Suppose a certain collection of termites is growing according to the function

$$f(t) = 3000e^{.04t},$$

where t is measured in months. If there are 3000 present on January 1, how many will be present on July 1?

62. Suppose the population of a small town is approximated by the function

$$p(t) = 10,000e^{.05t},$$

where t represents the time in years. The population at time $t = 0$ was 10,000. Find the population to the nearest thousand at time $t = 14$.

11.6 Exponential and Logarithmic Equations; Further Applications

Objective 1 Solve equations involving variables in the exponents.

Solve each equation. Give solutions to three decimal places.

1. $27^x = 5$

2. $32^y = 6$

3. $7^m = 11$

4. $12^{-p} = 32$

5. $4^{m-3} = 6$

6. $8^{4-x} = 3$

7. $2^{3y-9} = 7$

8. $4^{p+3} = 10$

9. $6^{2-r} = 50$

10. $5^{2k-1} = 17$

Objective 2 Solve equations involving logarithms.

Solve each equation. Give the exact solution.

11. $\log(p-2) = \log 3$

12. $\log(2k+1) = \log 7$

13. $\log_2(x+1) - \log_2 x = \log_2 5$

14. $\log_3(x-1) + \log_3 x = \log_3 6$

15. $\log(-y) + \log 4 = \log(2y+5)$

16. $\log_m 8 = 3$

17. $\log_p 10 = 4$

18. $\log_a 25 = \dfrac{1}{2}$

19. $\log_3 a = \log_3(a-1) + 2$

20. $\log_4 u = 1 - \log_4(u+3)$

Objective 3 Solve applications of compound interest.

Solve each problem.

Find the final amount owed for each of the following borrowed amounts if interest is compounded annually. Use

$$A = P\left(1 + \frac{r}{n}\right)^{nt},$$

where A is the amount owed, P is the amount borrowed, r is the interest rate, $n = 1$, and t is the time in years.

21. $1000 for 3 years at 8%

22. $25,000 for 5 years at 10%

23. $5600 for 8 years at 11%

24. $2700 for 10 years at 9%

25. $3950 for 5 years at 7%

26. $47,200 for 9 years at 10%

Objective 4   **Solve applications involving base e exponential growth and decay.**

Solve each problem.

27. Radioactive strontium decays according to the function
$$y = y_0 e^{-.0239t},$$
where t is the time in years. If an initial sample contains $y_0 = 15$ g of radioactive strontium, how many grams will be present after 25 years? Round to the nearest hundredth of a gram.

28. What is the half-life of radioactive strontium in Exercise 27? Round to the nearest year.

29. A sample of 500 g of lead-210 decays to polonium-210 according to the function
$$A(t) = 500e^{-.032t},$$
where t is the time in years. How much lead will be left in the sample after 20 years? Round to the nearest gram.

30. What is the half-life of the initial sample of lead in Exercise 29? Round to the nearest year.

Objective 5   **Use the change-of-base rule.**

Use the change-of-base rule to find each logarithm. Give approximations to four decimal places.

31. $\log_6 3$

32. $\log_2 10$

33. $\log_7 28$

34. $\log_5 180$

35. $\log_3 142$

36. $\log_{16} 27$

37. $\log_2 14$

38. $\log_5 243$

39. $\log_{1/2} 6$

40. $\log_{1/4} 11$

41. $\log_{1/7} 12$

42. $\log_{2/3} 5$

43. $\log_{1/2} \dfrac{3}{4}$

44. $\log_{1/3} \dfrac{1}{6}$

11.6 Mixed Exercises

Solve each equation. Give solutions to three decimal places.

45. $5^x = 16$

46. $3^{-q} = 2$

47. $10^{k-2} = 24$

48. $11^{5-w} = 7$

49. $\log_t 100 = \dfrac{2}{3}$

50. $\log_2 x + \log_2 (3x-1) = 1$

Find the final amount owed for each of the following borrowed amounts if interest is compounded annually.

51. $72,600 for 4 years at 12%

52. $32,800 for 7 years at 11%

Suppose that over several years a certain average annual rate of inflation is compounded annually. The formula for the price of an item can be found using the formula

$$A = P(1+r)^t,$$

where r is the rate of inflation and A is the price in dollars of the item after t years, if it cost P dollars when $t = 0$. Find the number of years, to the nearest year, that it takes an item to double in price for the following rates of inflation.

53. 4%   **54.** 11%

Use the change-of-base rule to find each logarithm. Give approximations to four decimal places.

55. $\log_8 12$ **56.** $\log_4 8$

Chapter 12

NONLINEAR FUNCTIONS, CONIC SECTIONS, AND NONLINEAR SYSTEMS

12.1 Additional Graphs of Functions; Composition

Objective 1 Recognize the graphs of the elementary functions defined by $|x|$, $\frac{1}{x}$, and $\sqrt{x}$, and graph their translations.

Graph each function.

1. $f(x) = |x-2| + 3$

2. $f(x) = \sqrt{x+3}$

3. $f(x) = \dfrac{1}{x-1}$

4. $f(x) = -|x+3| - 2$

5. $f(x) = \sqrt{5-x}$

6. $f(x) = \dfrac{1}{x} + 3$

7. $f(x) = |x-3| - 2$

8. $f(x) = \sqrt{x} + 3$

Objective 2 Recognize and graph step functions

Objective 3 Find the composition of functions.

Let $f(x) = x^2 + 3$, $g(x) = 3x + 2$, *and* $h(x) = x + 4$. *Find each composite function.*

9. $(h \circ g)(2)$

10. $(f \circ g)(1)$

11. $(g \circ f)(3)$

12. $(h \circ f)(4)$

13. $(f \circ h)(-3)$

14. $(f \circ g)(x)$

15. $(g \circ h)(x)$

16. $(f \circ h)(x)$

17. $(g \circ f)(x)$

18. $(h \circ g)(x)$

12.1 Mixed Exercises

Graph each function.

19. $f(x) = |x-4| + 2$

20. $f(x) = -\sqrt{x-3} - 3$

21. $f(x) = \dfrac{1}{x-3}$

22. $f(x) = -|x-3| + 3$

23. $f(x) = \sqrt{4-x^2}$

24. $f(x) = -\sqrt{1-x^2}$

Let $f(x) = x - 3$ *and* $g(x) = x^2 + 6$. *Find each composite function.*

25. $(f \circ g)(2)$

26. $(g \circ f)(5)$

27. $(f \circ g)(x)$

28. $(g \circ f)(x)$

12.2   The Circle and the Ellipse

Objective 1   **Find the equation of a circle given the center and radius.**

Find the equation of a circle with the given conditions.

1. center: $(-3, 2)$; radius: 5

2. center: $(1, 4)$; radius: 2

3. center: $(0, 5)$; radius: 3

4. center: $(6, 2)$; radius: 3

5. center: $(-5, 4)$; radius: 4

6. center: $(7, 1)$; radius: 2

7. center: $(3, -4)$; radius: 5

8. center: $(2, 2)$; radius: 6

9. center: $(1, 3)$; radius: 5

10. center: $(-2, -2)$; radius: 3

Objective 2   **Determine the center and radius of a circle given its equation.**

Find the center and radius of each circle. In Exercises 11 and 12, sketch each graph.

11. $x^2 + y^2 - 4x + 8y + 11 = 0$

12. $x^2 + y^2 + 6x - 4y + 12 = 0$

13. $x^2 + y^2 - 6x + 10y = 30$

14. $x^2 + y^2 - 4x - 2y = 31$

15. $x^2 + y^2 + 4x + 6y - 3 = 0$

16. $x^2 + y^2 - 10x + 12y + 52 = 0$

17. $x^2 + y^2 - 8x - 2y + 15 = 0$

18. $x^2 + y^2 - 4x + 8y + 11 = 0$

19. $2x^2 + 2y^2 + 4y - 8x = 4$

20. $3x^2 + 3y^2 + 12y + 30x = 21$

Objective 3   **Recognize the equation of an ellipse.**

Objective 4   **Graph ellipses.**

Graph each ellipse.

21. $\dfrac{x^2}{9} + \dfrac{y^2}{49} = 1$

22. $\dfrac{x^2}{25} + \dfrac{y^2}{4} = 1$

23. $\dfrac{x^2}{25} + \dfrac{y^2}{36} = 1$

24. $\dfrac{x^2}{4} + \dfrac{y^2}{9} = 1$

25. $\dfrac{x^2}{16} + \dfrac{y^2}{25} = 1$

26. $\dfrac{x^2}{36} + \dfrac{y^2}{9} = 1$

27. $\dfrac{x^2}{25} + \dfrac{y^2}{64} = 1$

28. $\dfrac{x^2}{4} + \dfrac{y^2}{16} = 1$

12.2 Mixed Exercises

Find the equation of a circle with the given conditions.

29. center: $(-2, -4)$; radius: 5

30. center: $(0, 3)$; radius: $\sqrt{2}$

Find the center and radius of each circle.

31. $x^2 + y^2 + 8x + 4y - 29 = 0$

32. $4x^2 + 4y^2 - 24x + 16y + 43 = 0$

Graph each ellipse.

33. $\dfrac{x^2}{16} + \dfrac{y^2}{49} = 1$

34. $\dfrac{x^2}{25} + \dfrac{y^2}{81} = 1$

12.3   The Hyperbola and Other Functions Defined by Radicals

Objective 1   **Recognize the equation of a hyperbola.**

Objective 2   **Graph hyperbolas by using the asymptotes.**

Graph each hyperbola.

1. $\dfrac{x^2}{9} - \dfrac{y^2}{16} = 1$

3. $\dfrac{y^2}{4} - \dfrac{x^2}{9} = 1$

5. $\dfrac{x^2}{36} - \dfrac{y^2}{49} = 1$

7. $\dfrac{x^2}{25} - \dfrac{y^2}{4} = 1$

2. $\dfrac{x^2}{25} - \dfrac{y^2}{9} = 1$

4. $\dfrac{y^2}{25} - \dfrac{x^2}{16} = 1$

6. $\dfrac{y^2}{4} - \dfrac{x^2}{4} = 1$

8. $\dfrac{x^2}{25} - \dfrac{y^2}{81} = 1$

Objective 3   **Identify conic sections by their equations.**

Identify each of the following as the equation of a parabola, a circle, an ellipse, or a hyperbola.

9. $x^2 = y^2 + 9$

13. $25y^2 + 100 = 4x^2$

17. $5x^2 = 25 - 5y^2$

10. $2x^2 + 3y^2 = 6$

14. $16x^2 + 9y = 144$

18. $y^2 = 36 - 36x^2$

11. $2x + y^2 = 16$

15. $16x^2 + 16y^2 = 64$

12. $4x^2 - 9y^2 = 36$

16. $2x^2 + 2y^2 = 8$

Objective 4   **Graph certain square root functions.**

Sketch each graph.

19. $f(x) = \sqrt{36 - x^2}$

23. $f(x) = \sqrt{1 + \dfrac{x^2}{4}}$

26. $f(x) = -5\sqrt{1 - \dfrac{x^2}{9}}$

20. $f(x) = \sqrt{25 - x^2}$

21. $f(x) = -\sqrt{4 - x^2}$

24. $f(x) = -3\sqrt{1 + \dfrac{x^2}{25}}$

22. $f(x) = -\sqrt{9 - x^2}$

25. $f(x) = \sqrt{9 - 9x^2}$

12.3 Mixed Exercises

Identify each of the following as the equation of a parabola, a circle, an ellipse, or a hyperbola.

27. $3x^2 - 3y = 9$

29. $x^2 = 49 - y^2$

31. $3x^2 = 3y^2 + 1$

28. $3x^2 + 3y^2 = 1$

30. $x^2 = 16 - y$

32. $x + y^2 = 16$

12.4 Nonlinear Systems of Equations

Objective 1 Solve a nonlinear system by substitution.

Solve each system by the substitution method.

1. $x^2 + y^2 = 17$
$2x = y + 9$

2. $2x^2 - y^2 = -1$
$2x + y = 7$

3. $4x^2 + 3y^2 = 7$
$2x - 5y = -7$

4. $x^2 = 2y^2 + 2$
$y = 3x + 7$

5. $y = x^2 - 3x - 8$
$x = y + 3$

6. $x = y^2 + 5y$
$3y = x$

7. $xy = -6$
$x + y = 1$

8. $xy = 24$
$y = 2x + 2$

9. $xy = -10$
$2x - y = 9$

10. $xy = 10$
$x + y = 7$

Objective 2 Use the elimination method to solve a system with two second-degree equations.

Solve each system by the the elimination method.

11. $x^2 + y^2 = 10$
$2x^2 - y^2 = -7$

12. $2x^2 + y^2 = 54$
$x^2 - 3y^2 = 13$

13. $2x^2 - 3y^2 = -19$
$4x^2 + y^2 = 25$

14. $x^2 - y^2 = 3$
$2x^2 + y^2 = 9$

15. $x^2 + 2y^2 = 11$
$2x^2 - y^2 = 17$

16. $3x^2 + 2y^2 = 30$
$2x^2 + y^2 = 17$

17. $5x^2 + y^2 = 6$
$2x^2 - 3y^2 = -1$

18. $4x^2 - 3y^2 = -8$
$2x^2 + y^2 = 5$

19. $3x^2 - 3y^2 = 9$
$4x^2 + y^2 = 17$

20. $3x^2 - 2y^2 = 12$
$x^2 + 3y^2 = 4$

Objective 3 Solve a system that requires a combination of methods.

Solve each system.

21. $x^2 + xy + y^2 = 43$
$x^2 + 2xy + y^2 = 49$

22. $x^2 + xy - y^2 = 5$
$-x^2 + xy + y^2 = -1$

23. $x^2 + 2xy + 3y^2 = 6$
$x^2 + 4xy + 3y^2 = 8$

24. $2x^2 + 3xy - 2y^2 = 50$
$x^2 - 4xy - y^2 = -41$

25. $4x^2 - 2xy + 4y^2 = 64$
$\quad\quad x^2 \quad\quad + \; y^2 = 13$

28. $\quad x^2 + 3xy + 2y^2 = 12$
$\quad\quad -x^2 + 8xy - 2y^2 = 10$

26. $5x^2 - xy + 5y^2 = 89$
$\quad\quad x^2 + \quad\quad y^2 = 17$

29. $x^2 + 5xy - y^2 = 20$
$\quad\quad x^2 - 2xy - y^2 = -8$

27. $\quad 3x^2 - 4xy + 2y^2 = \;\; 59$
$\quad\quad -3x^2 + 5xy - 2y^2 = -65$

30. $\quad 2x^2 + \;\; xy + y^2 = \;\; 16$
$\quad\quad -2x^2 + 3xy - y^2 = -28$

12.4 Mixed Exercises

Solve each system.

31. $x^2 + 3y^2 = 3$
$\quad x = 3y$

35. $y = x^2$
$\quad y = 2x^2 - x - 6$

39. $xy = 5$
$\quad 2x^2 - y^2 = 5$

32. $xy = 1$
$\quad x + 2y = 3$

36. $3x^2 + \;\; y^2 = 13$
$\quad 4x^2 - 3y^2 = 13$

40. $x^2 - xy + y^2 = 6$
$\quad x + y = 0$

33. $2x^2 + 3y^2 = 6$
$\quad\quad x^2 + 3y^2 = 3$

37. $\quad 3x^2 + 2xy - 3y^2 = 5$
$\quad\quad -x^2 - 3xy + \;\; y^2 = 3$

34. $x^2 + 2xy - y^2 = 7$
$\quad x^2 \quad\quad - y^2 = 3$

38. $x^2 - 3x + y^2 = 4$
$\quad\quad 2x - y \;\; = 3$

12.5 Second-Degree Inequalities and Systems of Inequalities

Objective 1 Graph second-degree inequalities.

Graph each inequality.

1. $x \geq y^2$

2. $y^2 \geq 9 - x^2$

3. $16x^2 < 9y^2 + 144$

4. $25y^2 \leq 100 - 4x^2$

5. $x^2 + 4y^2 > 4$

6. $y \geq x^2 - 4$

7. $x \leq 2y^2 + 8y + 9$

8. $4y^2 \geq 196 + 49x^2$

Objective 2 Graph the solution set of a system of inequalities.

Graph each system of inequalities.

9. $-x + y > 2$
   $3x + y > 6$

10. $x + y > -2$
    $2x - y \leq -4$

11. $x - 2y \geq -6$
    $x + 4y \geq 12$

12. $x^2 + y^2 \leq 25$
    $3x - 5y > -15$

13. $9x^2 + 16y^2 < 144$
    $y^2 - x^2 > 4$

14. $x^2 + y^2 \leq 16$
    $y \leq x^2 - 4$

12.5 Mixed Exercises

Graph each inequality or system of inequalities.

15. $7x^2 \leq 42 - 6y^2$

16. $9x^2 + 64y^2 \leq 576$
    $x \geq 0$

17. $x^2 > 9 - y^2$
    $x \leq 0$
    $y \geq 0$

18. $x^2 - y^2 \leq 16$
    $y \geq 0$

19. $4y + x^2 < 0$
    $x \geq 0$

20. $x^2 + 4y^2 \leq 36$
    $-5 < x < 2$
    $y \geq 0$

ANSWERS TO
ADDITIONAL EXERCISES

Chapter 1

REVIEW OF THE REAL NUMBER SYSTEM

1.1 Basic Concepts

Objective 1

1. True

2. False

3. False

4. True

5. False

6. False

7. a, b, c, d, e

8. No elements

9. 7, 8, 9, 10

10. Ohio, Oklahoma, Oregon

11. Bush, Clinton, Bush

12. 1, 3, 5, 7

13. 5, 10, 15, 20, …

Objective 2

14.

15.

16.

17.

18.

19.

20.

21.

22.

23.

Objective 3

24. 5

25. $-4, 0, 5$

26. $-\sqrt{2}, \sqrt{7}$

27. $0, 5$

28. $-4, -\dfrac{1}{3}, 0, \dfrac{4}{5}, 5$

29. $-4, -\sqrt{2}, -\dfrac{1}{3}, 0, \dfrac{4}{5}, \sqrt{7}, 5$

30. True

31. False

32. False

33. False

34. True

35. True

36. False

37. True

Objective 4

38. 0

39. 2

40. −3

41. −2

42. π

43. $-\sqrt{2}$

44. .75

45. −2.5

46. $\dfrac{5}{2}$

47. $\dfrac{6}{5}$

48. $-\dfrac{10}{17}$

49. $-\dfrac{8}{9}$

Objective 5

50. 12

51. 7

52. −8

53. −8

54. 0

55. 5

56. 9

57. 3

58. 15

59. 4

60. 15

61. 5

Objective 6

62. False

63. True

64. True

65. False

66. True

67. False

68. True

69. True

70. True

71. $21 > 13$

72. $-8 < 2$

73. $-6 \le -3$

74. $0 \ge x$

75. False

76. True

77. True

78. False

79. False

80. False

81. False

82. True

83. True

84. True

1.1 Mixed Exercises

85. False

86. True

87. False

88. True

89. False

90. (a) $-\dfrac{4}{3}$

   (b) $\dfrac{4}{3}$

91. (a) $\sqrt{5}$

   (b) $\sqrt{5}$

92. (a) 0

   (b) 0

93. −8

94. 0

95. 16

96. 17

97. $-8 > -14$

98. $-2 < 2$

99. $-4 < a < 5$

100. $-11 \le 15$

101. $6 \ge 6$

102. $-3 \le t < -1$

103. $0 \le w \le 11$

1.2   Operations on Real Numbers

Objective 1

1. -12

2. -12.4

3. -2

4. 0

5. 1

6. -7

7. -5

8. 4

9. $-\dfrac{1}{8}$

10. $-\dfrac{3}{14}$

11. -1

12. $-\dfrac{2}{15}$

13. $-\dfrac{16}{33}$

14. $\dfrac{5}{24}$

15. $-\dfrac{4}{3}$

16. $-\dfrac{15}{22}$

17. -14.05

18. -7.96

19. 9

20. 0

21. -12

22. -12

23. -12

24. -16

25. 14

26. 5

Objective 2

27. -4

28. -4

29. -10

30. 1

31. 10

32. -3.78

33. -3.61

34. 1

35. $\dfrac{14}{15}$

36. -6

37. $-.6$

38. -6

39. 1

40. 0

41. 2

42. -5.2

43. 14

44. -10

Objective 3

45. -28

46. 15

47. -1.2

48. -54

49. $-.84$

50. 105

51. $-\dfrac{27}{28}$

52. $\dfrac{3}{10}$

53. $-\dfrac{21}{5}$

54. $-\dfrac{4}{7}$

55. $-\dfrac{18}{5}$

56. $\dfrac{3}{2}$

57. $-\dfrac{3}{2}$

58. 3.924

59. -7.536

Objective 4

60. $\dfrac{1}{13}$

61. $-\dfrac{1}{5}$

62. $\dfrac{1}{7}$

63. $-\dfrac{1}{3}$

64. 3

65. $-\dfrac{5}{3}$

66. $\dfrac{7}{8}$

67. $-\dfrac{18}{11}$

68. $\dfrac{12}{23}$

69. $\dfrac{6}{5}$

70. 50

71. $\dfrac{20}{7}$

Objective 5

72. $-\dfrac{1}{4}$

73. $-\dfrac{1}{2}$

74. 0

75. $-\dfrac{6}{7}$

76. $\dfrac{1}{4}$

77. $-\dfrac{1}{2}$

78. -7.9

79. 100

80. -800

81. 0

82. Not defined

83. -3

84. $\dfrac{21}{10}$

85. $-\dfrac{1}{4}$

86. $\dfrac{1}{8}$

1.2 Mixed Exercises

87. Sometimes true; for example, $5 + (-3) = 2$ is positive, but $1 + (-2) = -1$ is negative.

88. Sometimes true; for example, $-2 - 4 = -6$ is negative, but $5 - (-2) = 7$ is positive.

89. Sometimes true; for example, the sign of $4(3) = 12$ has the same sign in the product, but $-4(-3) = 12$ has a different sign in the product.

90. Always true

91. Never true

92. $\dfrac{4}{3}$

93. 3

94. $\dfrac{73}{105}$

95. $\dfrac{1}{8}$

96. 12

97. 3

98. -8

99. -1

100. 3

101. 1

1.3   Exponents, Roots, and Order of Operations

Objective 1

1. 27

2. 4

3. −25

4. −243

5. −243

6. −8

7. −16

8. $\dfrac{9}{25}$

9. $-\dfrac{64}{27}$

10. $\dfrac{125}{512}$

11. .01

12. .25

Objective 2

13. Exponent 4, base 3

14. Exponent 3, base 12

15. Exponent 3, base −5

16. Exponent 5, base x

17. Exponent 6, base y

18. Exponent 2, base x

19. Exponent 7, base 5

20. Exponent 5, base 8.1

21. Exponent 0, base 12

22. Exponent 2, base −6

23. Exponent 4, base $5r$

24. Exponent 5, base $7p$

Objective 3

25. 4

26. 11

27. −17

28. 0

29. $\dfrac{9}{2}$

30. 4

31. Not a real number

32. −3

33. Not a real number

34. .7

35. .9

36. −40

37. Not a real number

38. $-\dfrac{2}{7}$

39. 12

40. −.1

41. −.5

42. Not a real number

43. Not a real number

44. 100

45. Not a real number

Objective 4

46. −20

47. −24

48. $-\dfrac{1}{2}$

49. 64

50. −31

51. 47

52. $\dfrac{9}{2}$

53. 5

54. $-\dfrac{3}{2}$

55. 0

56. $49 + 4\sqrt{5}$

57. 20

58. −9

59. $\dfrac{16}{3}$

60. $\dfrac{35}{2}$

61. −1

Objective 5

62. 8

63. 21

64. −27

65. 24

66. 40

67. −8

68. −11

69. 8

70. 124

71. 35

72. $-\dfrac{4}{5}$

73. $-\dfrac{14}{15}$

74. $-\dfrac{34}{13}$

75. $-\dfrac{33}{13}$

1.3 Mixed Exercises

76. 1296

77. −1.331

78. 1.4

79. −144

80. −10

81. $-\dfrac{625}{81}$

82. 88

83. −59

84. $-\dfrac{2}{3}$

85. $\dfrac{23}{26}$

86. $-\dfrac{11}{75}$

1.4   Properties of Real Numbers

Objective 1

1. $4x - 12$

2. $7t - 35$

3. $-3z + 21$

4. $-x + 6$

5. $13k$

6. $13x$

7. $-4y$

8. $-9y$

9. Cannot be simplified

10. Cannot be simplified

Objective 2

11. (-7)

12. 4

13. 3

14. -1.5

15. 0

16. $\dfrac{2}{3}$

17. $\left(-\dfrac{1}{7}\right)$

18. 8

19. $\left(-\dfrac{1}{6}\right)$

20. 1

21. $\dfrac{1}{5}$

22. $\dfrac{5}{4}$

Objective 3

23. 7

24. -6

25. 0

26. 0

27. -2.5

28. 8

29. -12

30. 1

31. $\dfrac{2}{3}$

32. $-\dfrac{7}{3}$

Objective 4

33. Commutative

34. Commutative

35. Associative

36. Commutative

37. Associative

38. Associative

39. Commutative

40. Associative

41. Associative

42. Commutative

43. Commutative

44. Commutative

Objective 5

45. 0

46. 0

47. 0

48. Any real number

49. 0

50. 0

51. 0

52. 0

53. 0

54. 0

1.4 Mixed Exercises

55. $12r$; distributive

56. $-2a$; identity, distributive

57. $10k + 3$; commutative, associative, distributive

58. $9s - 12t$; distributive, commutative, associative

59. $4m - n$; distributive, associative, identity

60. 0; multiplication property of zero

61. $17p - 12$; distributive, commutative, associative

62. $7j - 8$; distributive, associative

63. $-6x$; distributive, associative, inverse

64. n; associative, inverse

65. $5p + 18$; distributive, commutative, associative

66. $2m$; distributive, commutative, associative, inverse

67. $5w$

68. $-3x + 17$

69. $-8 - 3d$

Chapter 2

LINEAR EQUATIONS AND APPLICATIONS

2.1   Linear Equations in One Variable

Objective 1

1. Yes	**3.** Yes	**5.** Yes	**7.** Yes	**9.** Yes
2. No	**4.** Yes	**6.** No	**8.** No	**10.** Yes

Objective 2

11. $\{-9\}$

12. $\{-3\}$

13. $\{-1\}$

14. $\varnothing$

15. $\left\{-\dfrac{2}{7}\right\}$

16. $\{-5\}$

17. $\{1\}$

18. $\{8\}$

19. $\left\{-\dfrac{8}{3}\right\}$

20. $\left\{\dfrac{3}{2}\right\}$

Objective 3

21. $\{10\}$

22. $\{-2\}$

23. $\left\{\dfrac{19}{5}\right\}$

24. $\left\{\dfrac{5}{9}\right\}$

25. $\{0\}$

26. $\{0\}$

27. $\{4\}$

28. $\{-8\}$

29. $\{-4\}$

30. $\left\{\dfrac{2}{5}\right\}$

31. $\left\{-\dfrac{3}{4}\right\}$

32. $\left\{\dfrac{24}{19}\right\}$

33. $\{-1\}$

34. $\left\{-\dfrac{163}{25}\right\}$

Objective 4

35. $\{51\}$

36. $\left\{-\dfrac{98}{5}\right\}$

37. $\{-10\}$

38. $\{6\}$

39. $\left\{-\dfrac{3}{2}\right\}$

40. $\{-14\}$

41. $\{28\}$

42. $\left\{-\dfrac{108}{5}\right\}$

43. $\{2\}$

44. $\left\{-\dfrac{3}{2}\right\}$

45. $\{-5\}$

46. $\{3\}$

47. $\{300\}$

48. $\{60\}$

49. $\{210\}$

50. $\{21\}$

51. $\{47\}$

52. $\{146\}$

Objective 5

53. Conditional equation; $\{2\}$

54. Conditional equation; $\{6\}$

55. Contradiction; $\varnothing$

56. Identity; $\{$All real numbers$\}$

57. Contradiction; $\varnothing$

58. Identity; $\{$All real numbers$\}$

59. Conditional equation; $\{0\}$

60. Contradiction; $\varnothing$

61. Identity; $\{$All real numbers$\}$

62. Conditional equation; $\{1\}$

2.1 Mixed Exercises

63. No

64. Yes

65. $\{-3\}$

66. $\{11\}$

67. $\{-7\}$

68. $\left\{-\dfrac{5}{2}\right\}$

69. $\left\{\dfrac{14}{5}\right\}$

70. {All real numbers}

71. $\{12\}$

72. $\{-21\}$

73. $\{0\}$

74. $\left\{-\dfrac{24}{31}\right\}$

75. $\varnothing$

76. {All real numbers}

77. $\{8\}$

78. $\{6\}$

2.2   Formulas

Objective 1

1. $r = \dfrac{I}{pt}$

2. $B = \dfrac{3V}{h}$

3. $r = \dfrac{d}{2}$

4. $L = \dfrac{A}{W}$

5. $s = \dfrac{P}{4}$

6. $b = P - a - c$

7. $H = \dfrac{V}{WL}$

8. $h = \dfrac{V}{\pi r^2}$

9. $F = \dfrac{9C}{5} + 32$

10. $C = \dfrac{5(F - 32)}{9}$

11. $y = \dfrac{7}{2(1 - x)}$

12. $x = \dfrac{7}{r - 1}$

13. $y = \dfrac{2x}{3 - x}$

14. $x = \dfrac{6}{2y - 1}$

Objective 2

15. 3 min

16. 4 ft

17. 16

18. 6 ft

19. 150 km/hr

20. 4 in.

21. 8%

22. 40°C

23. 18 m

24. 9.8%

25. 10%

26. 2 in.

Objective 3

27. 20%

28. 25%

29. 37.5%

30. 40%

31. $1800

32. 10,000

33. 2.3%

34. $2500

35. 60%

36. 1100

37. 2.6%

38. 6.5%

39. 3.5%

40. $650

41. $1200

2.2 Mixed Exercises

42. $C = 180° - A - B$

43. $d = \dfrac{C}{\pi}$

44. 4 hr

45. $16\dfrac{2}{3}\%$

46. 8 oz

47. 52 mph

48. 10.4 m

49. 380 ft

50. 5°C

51. $28,000

52. $600

2.3 Applications of Linear Equations

Objective 1

1. $x + 13$

2. $x - (-5)$

3. $7x$

4. $x - 8$

5. $x + 10$

6. $14 - 2x$

7. $\dfrac{x}{5}$

8. $\dfrac{x}{-3}$

9. $-7 + 4x$

10. $\dfrac{x+4}{9}$

Objective 2

11. $x + 9 = 45$

12. $2x - 3 = 25$

13. $6x = 7 + 5x$

14. $\dfrac{x}{7} = 6$

15. $x + 2x = 12$

16. $\dfrac{x}{x-3} = 17$

17. $2x = 3(x + 7)$

18. $27 = 4 + 7x$

19. $.40(x + 3) = 5$

20. $18 - x = 4x$

Objective 3

21. Expression

22. Equation

23. Expression

24. Equation

25. Equation

26. Expression

Objective 4

27. Length: 16 ft; width: 14 ft

28. Length: 25 m; width: 6 m

29. 6 m, 7 m, 9 m

30. 8 ft, 8 ft, 10 ft

31. Length: 13 m; perimeter: 39 m

32. 12 ft

33. 312 cm

34. Length: 7 ft; width: 4 ft

35. 4860 walleyes, 6690 bass

36. Winning candidate: 1926 votes; losing candidate: 1289 votes

Objective 5

37. 350

38. 45

39. 650

40. $35

41. $5000

42. $480

43. 30%

44. 5%

45. $6370.50

46. $640

Objective 6

47. $1600 at 5%, $3500 at 7%

48. $4220 at 8%, $3520 at 7%

49. $200 at 8%, $700 at 6%

50. $300 at 9%, $800 at 7%

51. $8000

52. $2500 at 5.5%, $3000 at 4%

53. $6000

54. $64,000

Objective 7

55. 20 L

56. 0.5 L

57. 15 ml

58. 5 gal

59. 36 pt

60. 60 lb

61. 8 oz

2.3 Mixed Exercises

62. $2x + 50 = x - 6$

63. $\dfrac{x}{4} + 2x = 8$

64. $30x = x + 87$

65. 50%

66. 10 in. and 27 in.

67. 7.5 gal

68. $250,000

69. $25,080

70. $6000

71. 8 lb

72. $700 at 6%, $600 at 9%

2.4   Further Applications of Linear Equations

Objective 1

1. 28 nickels, 13 dimes

2. 32 quarters, 18 nickels

3. 35 pennies, 29 nickels, 30 dimes

4. 170

5. 50 $.19-stamps, 50 $.29-stamps, and 10 $2.90-stamps

6. 25 large jars, 55 small jars

Objective 2

7. 9 mph

8. $4\frac{1}{2}$ hr

9. $7\frac{1}{4}$ hr

10. 12 hr

11. 30 mi

12. 60 mph

13. 16 mph

14. 65 mph, 85 mph

15. 8 mph, 10 mph

16. 135 mph, 110 mph

Objective 3

17. 65°, 70°, 45°

18. 20°, 115°, 45°

19. 55°, 25° 100°

20. 60°, 30°, 90°

21. 165°, 15°

22. 142°, 38°

23. 70°, 20°

2.4 Mixed Exercises

24. 16 nickels, 14 pennies

25. 2 hr

26. 90°, 60°, 30°

27. $2\frac{3}{4}$ hr

28. 112 senior citizens, 224 children, 324 adults

29. 2 nickels, 26 dimes

30. 126°, 54°

31. $\frac{1}{2}$ hr

32. 50 mph, 40 mph

33. $\frac{5}{6}$ hr or 50 min

Chapter 3

LINEAR EQUATIONS AND INEQUALITIES

3.1   Linear Inequalities in One Variable

Objective 1

1. $(-4, \infty)$

2. $(-\infty, 6)$

3. $[2, \infty)$

4. $(-\infty, 0]$

5. $[-2, 3]$

6. $(1, 5]$

7. $[-4, 0)$

8. $(-5, -1)$

Objective 2

9. $(-\infty, -3]$

10. $(9, \infty)$

11. $(-6, \infty)$

12. $[9, \infty)$

13. $(-\infty, 0)$

14. $(3, \infty)$

15. $(-\infty, -3)$

16. $[-2, \infty)$

17. $(1, \infty)$

18. $(-\infty, 15]$

19. $(4, \infty)$

20. $(6, \infty)$

Objective 3

21. $(-\infty, 3]$

22. $(-3, \infty)$

23. $(-\infty, -4)$

24. $\left(-\infty, -\dfrac{5}{2}\right]$

25. $[4, \infty)$

26. $(-\infty, -10]$

27. $(-\infty, -36]$

28. $[10, \infty)$

29. $(-\infty, 14)$

30. $(-\infty, -2]$

31. $(-\infty, -4)$

32. $(-8, \infty)$

33. $[-3, \infty)$

Objective 4

34. $(1, 9)$

35. $(-8, 6)$

36. $(5, 9)$

37. $[-9, -7)$

38. $(4, 5)$

39. $\left(-\dfrac{8}{3}, 0\right)$

40. $\left[-\dfrac{19}{2}, \dfrac{29}{2}\right]$

41. $\left[-\dfrac{11}{6}, \dfrac{1}{6}\right]$

42. $[-8, 12]$

43. $[-15, 10]$

Objective 5

44. The number is between −2 and 2.

45. The number is between −10 and 6.

46. The number is less than 3.

47. The number is greater than or equal to 10.

48. The number is less than or equal to 8.

49. The number is less than or equal to 1.

50. The number is less than $\frac{7}{6}$.

51. $\leq$

52. $\geq$

53. $\geq$

54. $\leq$

55. At least 5 dimes

56. 84 points

57. 4 summers

3.1 Mixed Exercises

58. $\left(-\dfrac{3}{2}, \infty\right)$

59. $\left[\dfrac{1}{2}, \dfrac{25}{2}\right]$

60. $\left(-\dfrac{1}{3}, \dfrac{2}{3}\right)$

$-\dfrac{1}{3}$ $\dfrac{2}{3}$

61. $\left(-\dfrac{7}{2}, 1\right)$

$-\dfrac{7}{2}$ 1

62. $\left[-1, \dfrac{13}{3}\right]$

-1 $\dfrac{13}{3}$

63. $\left(-\infty, -\dfrac{3}{8}\right]$

$-\dfrac{3}{8}$

64. At least 74

65. At least 5 red lights

66. At least 4 yellow pills

67. More than 6 pots

3.2   Set Operations and Compound Inequalities

Objective 1

1. $\{2, 3\}$ **3.** $\varnothing$ **5.** $\{2, 4\}$ **7.** $\{0, 2, 4\}$ or D **9.** $\varnothing$

2. $\varnothing$ **4.** $\{2, 6, 8\}$ **6.** $\{1, 3, 5\}$ **8.** $\{0\}$ or E **10.** $\{2, 4\}$

Objective 2

11. $(0, 3)$

12. $(-\infty, 4]$

13. $[1, \infty)$

14. $[5, 9]$

15. $(-2, -1)$

16. $\varnothing$

17. $[-4, 3)$

18. $(0, 1]$

19. $\varnothing$

20. $\varnothing$

Objective 3

21. $\{0, 1, 2, 3, 4, 5\}$

22. $\{-6, -5, -4, -3, -2, -1\}$

23. $\{7, 8, 9, 10\}$

24. $\{2, 6, 8\}$

25. $\{0, 1, 2, 3, 4, 5, 6, 8, 10\}$

26. $\{1, 2, 3, 4, 5, 6, 7, 9\}$

27. $\{1, 2, 3, 4, 5, 6\}$ or A

28. $\{0, 1, 2, 3, 4, 5, 6\}$

29. $\{0, 1, 2, 3, 4, 5, 6, 7, 8, 9, 10\}$

30. $\{0, 1, 2, 3, 4, 6, 8, 10\}$

Objective 4

31. $(-\infty, -1) \cup (4, \infty)$

32. $(-\infty, 1] \cup [6, \infty)$

33. $(-\infty, 2] \cup [6, \infty)$

34. $(-\infty, 3] \cup [6, \infty)$

35. $[-1, \infty)$

36. $(-\infty, \infty)$

37. $(-\infty, -4] \cup (4, \infty)$

38. $(-\infty, -1) \cup (5, \infty)$

39. $\left(-\infty, -\dfrac{5}{3}\right) \cup \left(\dfrac{3}{4}, \infty\right)$

3.2 Mixed Exercises

41. $\{0, 2, 4\}$

42. $\{1, 3, 5\}$

43. $\{0, 1, 2, 3, 4, 5\}$ or D

44. $\varnothing$

45. $(-\infty, -2] \cup [0, \infty)$

46. $[-2, 3)$

47. $(-\infty, 1) \cup [2, \infty)$

40. $(-\infty, \infty)$

48. $(-\infty, 5) \cup (6, \infty)$

49. $\varnothing$

50. $(-\infty, \infty)$

51. $(-1, 4]$

52. $\left(-\infty, -\dfrac{2}{7}\right]$

3.3   Absolute Value Equations and Inequalities

Objective 1

1.

2.

3.

4.

5.

6.

7.

8.

9.

10.

Objective 2

11. $\{-3, 11\}$

12. $\{-12, 4\}$

13. $\{-4, 10\}$

14. $\left\{-\dfrac{5}{3}, \dfrac{7}{3}\right\}$

15. $\{2, 8\}$

16. $\{-8, -4\}$

17. $\left\{-\dfrac{13}{2}, \dfrac{7}{2}\right\}$

18. $\left\{-\dfrac{3}{2}\right\}$

19. $\{3\}$

20. $\left\{-3, \dfrac{21}{2}\right\}$

Objective 3

21. $(-\infty, -6) \cup (10, \infty)$

22. $(-\infty, -20) \cup (10, \infty)$

23. $(-13, 3)$

24. $(-3, 13)$

25. $\left(-4, \dfrac{16}{5}\right)$

26. $(-\infty, -7] \cup [16, \infty)$

27. $\left(-\infty, \dfrac{1}{3}\right] \cup [3, \infty)$

28. $(-\infty, -4] \cup [-2, \infty)$

29. $(-\infty, -3) \cup (-1, \infty)$

30. $(-\infty, 0] \cup [10, \infty)$

31. $[-1, 5]$

32. $\left[-\dfrac{1}{4}, \dfrac{3}{4}\right]$

Objective 4

33. $\{-2, 2\}$

34. $\{-9, 9\}$

35. $\varnothing$

36. $\varnothing$

37. $\{-9, -1\}$

38. $\left\{-\dfrac{13}{7}, \dfrac{3}{7}\right\}$

39. $\varnothing$

40. $\{-2, 3\}$

41. $\{-42, 50\}$

42. $\left\{-\dfrac{5}{4}, -\dfrac{1}{4}\right\}$

Objective 5

43. $\left\{\dfrac{3}{5}, 15\right\}$

44. $\left\{\dfrac{7}{2}\right\}$

45. $\left\{\dfrac{2}{3}, 8\right\}$

46. $\left\{-8, \dfrac{2}{3}\right\}$

47. $\{-1\}$

48. $\left\{-\dfrac{3}{2}, 2\right\}$

49. $\left\{-\dfrac{15}{4}, \dfrac{1}{8}\right\}$

50. $\left\{-3, \dfrac{11}{3}\right\}$

51. $\{-1, 1\}$

52. $\left\{-\dfrac{1}{2}, \dfrac{5}{2}\right\}$

Objective 6

53. $\varnothing$

54. $\{-14\}$

55. $\{0\}$

56. $\varnothing$

57. $\varnothing$

58. $\varnothing$

59. $\varnothing$

60. $\varnothing$

61. $(-\infty, \infty)$

62. $(-\infty, \infty)$

3.3 Mixed Exercises

63. $\{-4, -1\}$

64. $\{-2, 14\}$

65. $\varnothing$

66. $\{-11, 1\}$

67. $\varnothing$

68. $\left\{-\dfrac{29}{3}, \dfrac{25}{3}\right\}$

69. $\left\{-5, -\dfrac{3}{4}\right\}$

70. $\{-6, 0\}$

71. $(-2, 5)$

72. $\left[-\dfrac{1}{3}, \dfrac{19}{3}\right]$

73. $(-\infty, -8) \cup (1, \infty)$

74. $(-\infty, -4] \cup [10, \infty)$

75. $(-\infty, -6] \cup [5, \infty)$

76. $\left(-\infty, -\dfrac{1}{2}\right) \cup \left(\dfrac{1}{2}, \infty\right)$

77. $\left[-\dfrac{1}{3}, 1\right]$

78. $\left(-\dfrac{1}{2},\, 2\right)$

Chapter 4

GRAPHS, LINEAR EQUATIONS, AND FUNCTIONS

4.1 The Rectangular Coordinate System

Objective 1

1–8.

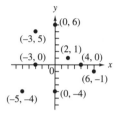

Objective 2

9. (0, 6), (6, 0), (2, 4), (1, 5)

10. (0, –4), (4, 0), (6, 2), (1, –3)

11. (0, 6), (3, 0), (4, –2), $\left(\dfrac{1}{2}, 5\right)$

12. (0, 6), (4, 0), (6, –3), $\left(5, -\dfrac{3}{2}\right)$

13. (0, –2), (8, 0), (–4, –3), (20, 3)

14. (0, –2), (5, 0), (–5, –4), $\left(-\dfrac{5}{2}, -3\right)$

15. (0, 7), (3, 0), (6, –7), (–3, 14)

16. (0, 8), (5, 0), (10, –8), $\left(\dfrac{15}{2}, -4\right)$

17. (–4, 0), (–4, –2), (–4, –5), (–4, 7)

18. (0, 2), (4, 2), (–1, 2), (–9, 2)

19. (0, –4), (–2, –4), (8, –4), (–4, –4)

20. (5, 0), (5, 9), (5, 5), (5, 2)

Objectives 3 and 4

21. (6, 0), (0, 6)

22. (–3, 0), (0, –3)

23. (1, 0), (0, –1)

24. (–7, 0), (0, 7)

25. (2, 0), (0, 3)

26. (2, 0), (0, 5)

27. (4, 0), (0, –5)

28. (3, 0), (0, –7)

29. (2, 0), $\left(0, \dfrac{11}{2}\right)$

30. $\left(\dfrac{18}{5}, 0\right)$, (0, 2)

31. $\left(-\dfrac{8}{3}, 0\right)$, $\left(0, -\dfrac{8}{7}\right)$

32. $\left(-\dfrac{5}{4}, 0\right)$, $\left(0, \dfrac{5}{7}\right)$

33. (2, 0), no *y*-intercept

34. no *x*-intercept, (0, 5)

35. no *x*-intercept, (0, –3)

36. (–4, 0), no *y*-intercept

37. (0, 0), (0, 0)

38. (0, 0), (0, 0)

39. $x + y = 6$

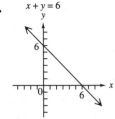

40. $x + y = -3$

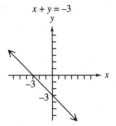

41. $x - y = 1$

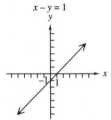

42. $x - y = -7$

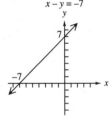

43. $3x + 2y = 6$

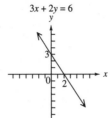

44. $5x + 2y = 10$

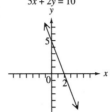

45. $5x - 4y = 20$

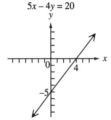

46. $7x - 3y = 21$

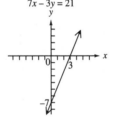

47. $4x + y = -6$

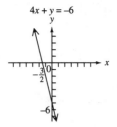

48. $3x - 2y = -12$

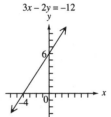

49. $4x - 3y = 0$

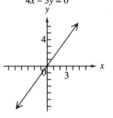

50. $3x + 2y = 0$

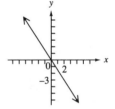

Objective 5

51. $x = 2$

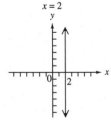

52. $y = 0$

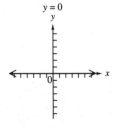

53. $y = -3$

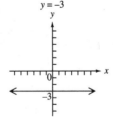

54. $x = -5$

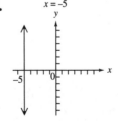

55. $y - 4 = 0$

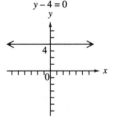

56. $x + 3 = 0$

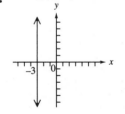

57. $y + 6 = 0$

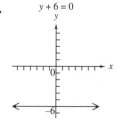

58. $x - 3 = 0$

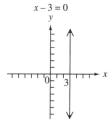

4.1 Mixed Exercises

59. $(1, 0), (0, -3), (2, 3), (4, 9)$

60. $\left(\dfrac{15}{2}, 0\right), (0, 3), (-5, 5), (5, 1)$

61. $(8, 0), (0, 2), \left(-1, \dfrac{9}{4}\right), (20, -3)$

62. $(0, 0), (0, 0), \left(4, \dfrac{8}{3}\right), \left(\dfrac{3}{2}, 1\right)$

63. $(4, 0), (0, -4)$
$x - y = 4$

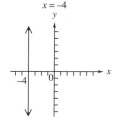

65. $(-4, 0)$, no y-intercept
$x = -4$

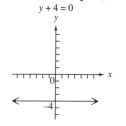

67.   no x-intercept, $(0, -4)$
$y + 4 = 0$

64. $(2, 0), (0, 6)$
$3x + y = 6$

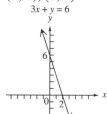

66. $(0, 0), (0, 0)$
$3x - y = 0$

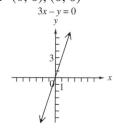

68. $(3, 0), \left(0, -\dfrac{9}{4}\right)$
$3x - 4y = 9$

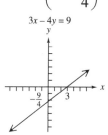

4.2 Slope

Objective 1

1. 1

2. 1

3. $\dfrac{3}{2}$

4. −1

5. 2

6. $\dfrac{5}{7}$

7. $\dfrac{5}{2}$

8. −3

9. $-\dfrac{1}{2}$

10. $\dfrac{1}{2}$

11. 0

12. 0

13. $\dfrac{1}{2}$

14. $-\dfrac{4}{5}$

Objective 2

15. 4

16. 2

17. −1

18. −3

19. −1

20. 0

21. $-\dfrac{4}{3}$

22. $\dfrac{3}{2}$

23. $-\dfrac{9}{5}$

24. $\dfrac{7}{3}$

25. $\dfrac{2}{7}$

26. $-\dfrac{1}{6}$

27. 0

28. undefined

29. 0

30. undefined

Objective 3

31.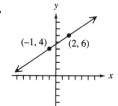

34.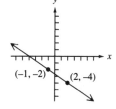

37.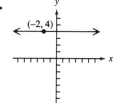

32.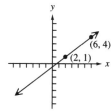

35.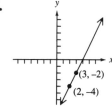

38.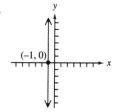

33.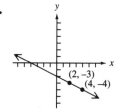

36.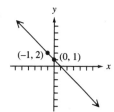

Objective 4

39. parallel

40. perpendicular

41. perpendicular

42. parallel

43. parallel

44. neither

45. neither

46. neither

47. neither

48. perpendicular

Objective 5

49. $1250/yr

50. 1500

51 **(a)** $7540/yr   **(b)** $6310

    (c) $6925/yr

52. 113.5 ft/min

53. .24 horizontal ft/1 vertical ft

54. **(a)** −350 students/yr

    (b) The enrollment was decreasing.

4.2 Mixed Exercises

55. $-\dfrac{1}{3}$

56. 0

57. $-\dfrac{2}{3}$

58. $\dfrac{5}{2}$

59. undefined

60. 0

61. −1

62. $\dfrac{1}{3}$

63.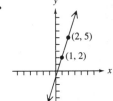

64.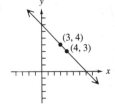

65.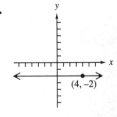

66.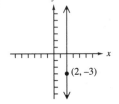

67. parallel

68. parallel

69. parallel

70. perpendicular

71. 5.4 employees/yr

72. 3,033,100

4.3   Linear Equations in Two Variables

Objective 1

1. $2x - y = 5$

2. $6x - y = 2$

3. $4x + y = 3$

4. $5x + y = 3$

5. $2x + 3y = 6$

6. $x + 4y = -12$

7. $3x - 5y = -2$

8. $6x - 5y = 1$

9. $7x - 3y = -27$

10. $y = 3$

Objective 3

11. $2x - y = -9$

12. $4x - y = 2$

13. $5x - y = 21$

14. $x - y = -9$

15. $5x + y = -1$

16. $3x + y = 3$

17. $x + 2y = 1$

18. $2x + 3y = -13$

19. $3x + 4y = -15$

20. $4x + 5y = 2$

21. $x = 3$

22. $y = 2$

23. $y = -4$

24. $x = 0$ (the y-axis)

25. $x = 2$

26. $y = 0$ (the x-axis)

Objective 4

27. $x - y = -5$

28. $4x + y = 29$

29. $3x + 2y = 23$

30. $x + 3y = -1$

31. $x + 2y = -2$

32. $7x + 4y = 13$

33. $x + 2y = -7$

34. $x + y = -5$

35. $y = 1$

36. $y = -5$

37. $x = 0$ (the y-axis)

38. $x = -1$

Objective 5

39. $x - y = 11$

40. $2x + 3y = 9$

41. $x + 3y = -5$

42. $4x - 3y = -17$

43. $5x + y = 4$

44. $3x + y = -28$

45. $x - 5y = 7$

46. $2x + 3y = 1$

47. $y = 7$

48. $x = 2$

49. $y = 6$

50. $x = -3$

Objective 6

51. **(a)** $y = .13x$

   (b) $(0, 0), (5, .65), (10, 1.30)$

52. **(a)** $y = 8x$

   (b) $(0, 0), (5, 40), (10, 80)$

53. **(a)** $y = 1.25x$

   (b) $(0, 0), (5, 6.25), (10, 12.50)$

54. **(a)** $y = .13x + .35$

   (b) $(0, .35), (5, 1.00), (10, 1.65)$

55. **(a)** $y = 8x + 15$

   (b) $(0, 15), (5, 55), (10, 95)$

56. **(a)** $y = 1.25x + 25$

   (b) $(0, 25), (5, 31.25), (10, 37.50)$

57. $1.91 = .13x + .35$; 12 min

58. $207 = 8x + 15$; 24 rows

59. $62.50 = 1.25x + 25$; 30 lines

4.3 Mixed Exercises

60. $15x + 24y = -16$

61. $x - y = -3$

62. $4x + 7y = -56$

63. $x = -3$

64. $3x + 4y = -8$

65. $y = -5$

66. $y = 2$

67. $x = -2$

68. $x - y = -3$

69. $3x + y = 25$

70. $3x + y = -1$

71. $y = 5$

72. $-\dfrac{2}{7}$; $(0, 2)$

73. $\dfrac{3}{2}$; $\left(0, -\dfrac{9}{2}\right)$

74. undefined slope; no y-intercept

75. 0; $(0, 2)$

4.4   Linear Inequalities in Two Variables

Objective 1

1. $x + y \geq 2$

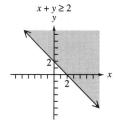

5. $2x + 3y \geq 6$

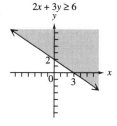

9. $x \leq 4y$

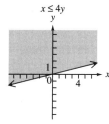

2. $x + y \leq 5$

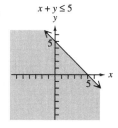

6. $3x - 2y \leq 12$

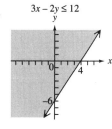

10. $y \geq -3$

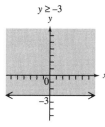

3. $x - y < 5$

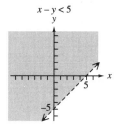

7. $3x - y \geq -3$

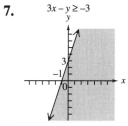

4. $x - y < -4$

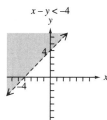

8. $2x - 5y < -10$

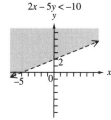

Objective 2

11. $x + y \leq 4$
and $x - y \geq 2$

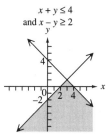

12. $x - y < 3$
and $x + y > -2$

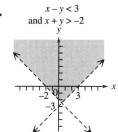

13. $2x + y < 6$
and $x - 3y > -6$

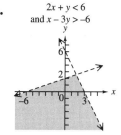

14. $4x + y \le 4$
and $x - 2y \le -2$

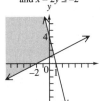

16. $5x + 2y < 10$
and $2x + 3y > 6$

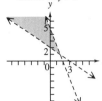

18. $4y - 3x \le 12$
and $x \ge 0$

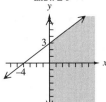

15. $3x + 4y \le 12$
and $2x - y \le 4$

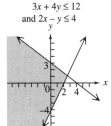

17. $2x + y \ge 6$
and $y \ge 2$

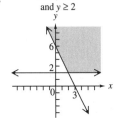

Objective 3

19. $4x - 2y \ge -4$
or $x \ge 1$

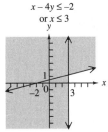

22. $2x + y < -1$
or $x - 2y > 1$

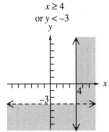

25. $x + y \ge 0$
or $x - y \ge 0$

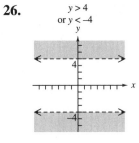

20. $x - 4y \le -2$
or $x \le 3$

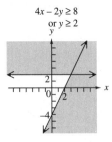

23. $x \ge 4$
or $y < -3$

26. $y > 4$
or $y < -4$

21. $4x - 2y \ge 8$
or $y \ge 2$

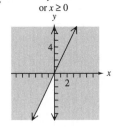

24. $y \ge 2x$
or $x \ge 0$

4.4 Mixed Exercises

27. $y + 4x < 0$

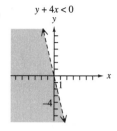

30. $y > 3x$

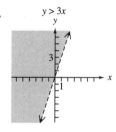

33. $x \leq -2$

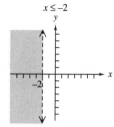

28. $x - y \leq 4$
and $x + y \geq -3$

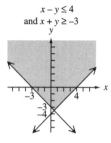

31. $x > 2$
and $2x - 3y < 6$

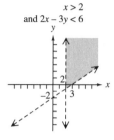

34. $x + y \geq 2$
or $2x - y \leq 4$

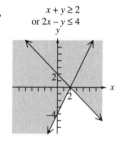

29. $x + 2y > 2$
or $y \leq 0$

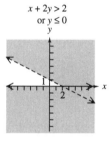

32. $x - 2y \geq 0$

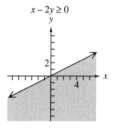

4.5 Introduction to Functions

Objective 1

1. not a function

2. function

3. function

4. function

5. not a function

6. function

7. not a function

8. not a function

9. function

10. function

11. not a function

12. not a function

13. function

14. function

15. function

Objective 2

16. function; domain: $\{-1, -2, 0\}$;
 range: $\{1, 2, 0\}$

17. function; domain: $\{3, 2, 1, -1\}$;
 range: $\{0, 4, 6, 3\}$

18. function; domain: $\{-2, -1, 0, 1\}$;
 range: $\{-2, -1, 0\}$

19. function; domain: $\{3, 2, 1\}$;
 range: $\{5, 3, 0\}$

20. not a function; domain: $\{1, 2, -1\}$;
 range: $\{3, -1, 4\}$

21. function; domain: $\{2, 1, -1, 0\}$;
 range: $\{-4, -2, 2, 3\}$

22. function; domain: $\{5, 3, 1, -1\}$;
 range: $\{2, -1, -3, -5\}$

23. function; domain: $\{4, 3, 2, 1, 0\}$;
 range: $\{2\}$

24. function; $(-\infty, \infty)$

25. not a function; $(-\infty, \infty)$

26. function; $[4, \infty)$

27. not a function; $[-1, \infty)$

28. function; $(-\infty, 0) \cup (0, \infty)$

29. function; $(-\infty, \infty)$

Objective 3

30. function

31. function

32. not a function

33. not a function

34. function

35. function

36. function

37. not a function

Objective 4

38. $1; 13; -2x + 5$

39. $10; -2; 6 + 2x$

40. $12; 48; 3x^2$

41. $8; 8; x^2 + 2x$

42. $\dfrac{4}{5}; \dfrac{4}{17}; \dfrac{4}{x^2+1}$

43. $-\dfrac{3}{5}; \dfrac{9}{5}; \dfrac{-2x+1}{5}$

Objective 5

44. $f(x) = x + 2$

45. $f(x) = 2x + 2$

46. $f(x) = x - 1$

47. $f(x) = -\dfrac{1}{2}x - 2$

48. $f(x) = -\dfrac{1}{4}x + \dfrac{1}{2}$

49. $f(x) = 3x$

50. $f(x) = -x + 1$

51. $f(x) = -\dfrac{3}{2}x - 3$

4.5 Mixed Exercises

52. function; domain: {0, 1, 2, 4}; range: {1, 3, –4, –8}

53. function; domain: {–2, –3, 1}; range: {–5, –2}

54. not a function; domain: {–4, –5, –3, –2}; range: {1, 2, 3, 4, 5}

55. function; domain: {1, 2, 3, 4, 5}; range: {10, 9, 8, 7, –4}

56. function; domain: $(-\infty, \infty)$

57. function; domain $[-2, \infty)$

58. function; domain: $(-\infty, \infty)$

59. not a function; domain: $[-4, \infty)$

60. $f(x) = -\dfrac{x}{3} + 3;\ \dfrac{10}{3};\ 1;\ -\dfrac{x}{3} + \dfrac{10}{3}$

61. $f(x) = 2x - 12;\ -14;\ 0;\ 2x - 14$

4.6 Variation

Objectives 1 and 2

1. $y = 3x$

2. $y = 5x$

3. $y = \dfrac{3x}{2}$

4. $y = \dfrac{x}{2}$

5. $y = \dfrac{23x}{12}$

6. $y = \dfrac{14x}{9}$

7. $y = 50x$

8. $y = 30x$

9. $y = \dfrac{4x}{3}$

10. $y = .25x$

11. 45

12. 144

13. $\dfrac{63}{2}$

14. $\dfrac{165}{4}$

15. 125

16. 147

17. 18

18. $\dfrac{49}{4}$

19. 69.08 cm

20. 100 psi

21. 100 newtons

22. 36π in.^{2}

Objective 3

23. $y = \dfrac{20}{x}$

24. $y = \dfrac{24}{x}$

25. $y = \dfrac{80}{x}$

26. $y = \dfrac{24}{x}$

27. $y = \dfrac{5}{6x}$

28. $y = \dfrac{27}{2x}$

29. $\dfrac{5}{2}$

30. 3

31. $\dfrac{4}{9}$

32. $\dfrac{64}{25}$

33. 100 amps

34. $\dfrac{400}{27}$ footcandles

35. 40 lb

36. 90 revolutions/min

Objective 4

37. $y = xz$

38. $y = \dfrac{25xz}{12}$

39. $y = \dfrac{xz}{2}$

40. $y = 2xz$

41. $y = \dfrac{xz}{2}$

42. $y = \dfrac{xz}{4}$

43. 24

44. 96

45. 128

46. 96

47. \$800

48. 750°

Objective 5

49. $y = \dfrac{3x}{z}$

50. $y = \dfrac{12x}{z}$

51. $y = \dfrac{2x}{3z}$

52. $y = \dfrac{x}{3z}$

53. $y = \dfrac{.65x}{z}$

54. $y = \dfrac{3.5x}{z}$

55. 9 hr

56. 6,000,000 dynes

4.6 Mixed Exercises

57. $p = 7q$

58. $w = .02vs$

59. $\dfrac{14}{3}$

60. 256.25

61. $\dfrac{128}{81}$

62. 120

63. 144 ft

64. 850 ohms

65. 1.105 L

66. 20 lb

67. 12 km

68. $810

Chapter 5

SYSTEMS OF LINEAR EQUATIONS

5.1   Systems of Linear Equations in Two Variables

Objective 1

1. $\{(1, 2)\}$

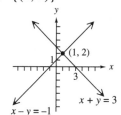

2. $\{(4, 1)\}$

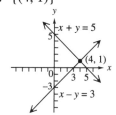

3. $\{(-2, 0)\}$

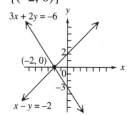

4. $\{(1, 0)\}$

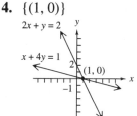

5. $\{(-2, 3)\}$

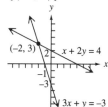

6. $\{(3, 1)\}$

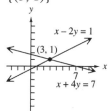

7. $\{(-4, 2)\}$

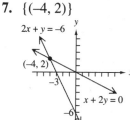

8. $\{(0, 2)\}$

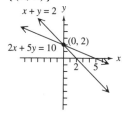

Objective 2

9. solution

10. solution

11. not a solution

12. not a solution

13. not a solution

14. solution

15. solution

16. solution

17. not a solution

18. not a solution

Objective 3

19. $\{(-4, -8)\}$

20. $\{(-1, 3)\}$

21. $\{(3, 1)\}$

22. $\{(3, -1)\}$

23. $\{(3, 1)\}$

24. $\{(-1, 8)\}$

25. $\{(3, 5)\}$

26. $\left\{\left(-\dfrac{1}{2}, 4\right)\right\}$

27. $\{(3, 2)\}$

28. $\{(-3, -4)\}$

29. $\{(1, 3)\}$

30. $\{(-1, -5)\}$

31. $\{(-1, 5)\}$

32. $\{(2, 1)\}$

33. $\{(3, 2)\}$

Objective 4

34. $\{(1, 3)\}$

35. $\{(4, 1)\}$

36. $\{(4, -2)\}$

37. $\{(-3, -1)\}$

38. $\{(-1, 7)\}$

39. $\left\{\left(\dfrac{1}{2}, \dfrac{1}{4}\right)\right\}$

40. $\left\{\left(\dfrac{2}{3}, \dfrac{1}{4}\right)\right\}$

41. $\{(6, 8)\}$

42. $\{(-5, 2)\}$

43. $\{(-2, -3)\}$

Objective 5

44. $\varnothing$

45. $\varnothing$

46. $\left\{(x, y) \mid 5x - 2y = 4\right\}$

47. $\left\{(x, y) \mid 9x - y = 6\right\}$

48. $\varnothing$

5.1   Mixed Exercises

49. $\{(3, 4)\}$

50. $\{(3, -1)\}$

51. $\{(0, -4)\}$

52. $\{(3, -1)\}$

53. $\left\{(x, y) \mid 4x + 6y = 24\right\}$

54. $\{(6, 2)\}$

55. $\{(2, -1)\}$

56. $\{(-3, 0)\}$

57. $\{(2, 4)\}$

58. $\varnothing$

59. $\varnothing$

60. $\left\{(x, y) \mid -x + 2y = -8\right\}$

5.2   Systems of Linear Equations in Three Variables

Objective 2

1. $\{(-2, 1, 1)\}$

2. $\{(1, 0, 3)\}$

3. $\{(3, -1, 2)\}$

4. $\{(-3, 1, 2)\}$

5. $\{(0, -2, 5)\}$

6. $\{(3, 0, -2)\}$

7. $\{(4, -4, 1)\}$

8. $\{(3, -6, 1)\}$

9. $\{(-12, 18, 0)\}$

10. $\{(-15, 0, 16)\}$

Objective 3

11. $\{(2, -1, 5)\}$

12. $\left\{\left(\dfrac{1}{2}, \dfrac{2}{3}, \dfrac{1}{5}\right)\right\}$

13. $\{(2, -5, 3)\}$

14. $\{(-1, -1, 5)\}$

15. $\{(3, 5, 0)\}$

16. $\{(0, -2, 5)\}$

17. $\{(4, 2, -1)\}$

18. $\{(2, -3, 1)\}$

19. $\{(6, -4, 1)\}$

20. $\{(-3, 5, -6)\}$

Objective 4

21. $\varnothing$

22. $\varnothing$

23. $\left\{(x, y, z)\,\middle|\,3x - 2y + 4z = 5\right\}$

24. $\left\{(x, y, z)\,\middle|\,-x + 5y - 2x = 3\right\}$

25. $\{(0, 0, 0)\}$

26. $\{(0, 0, 0)\}$

27. $\varnothing$

28. $\varnothing$

29. $\left\{(x, y, z)\,\middle|\,x - 5y + 2z = 0\right\}$

30. $\left\{(x, y, z)\,\middle|\,3x - 2y + 5z = 0\right\}$

5.2   Mixed Exercises

31. $\{(1, 2, 3)\}$

32. $\{(3, 2, 1)\}$

33. $\{(0, 5, -3)\}$

34. $\left\{(x, y, z)\,\middle|\,-3x - y + 2z = -3\right\}$

35. $\{(2, 2, 5)\}$

36. $\varnothing$

37. $\{(2, -1, 0)\}$

38. $\{(2, -1, 3)\}$

39. $\{(-2, -4, 0)\}$

40. $\{(1, 1, 2)\}$

5.3   Applications of Systems of Linear Equations

Objective 1

1. 12 ft

2. 30 cm

3. 15 in.

4. 21 in.

5. 76 yd

6. 80°

7. 12 cm

8. 21 cm

Objective 2

9. 20 fives, 20 tens

10. 15 tens, 25 twenties

11. large: $3; small: $2

12. small: $.30; large: $.50

13. marigold: $12.29; carnation: $17.60

14. cattle feed; $.75; rabbit feed: $.50

15. $7000 at 6%; $33,000 at 4%

16. $14,000 at 5%; $6000 at 6%

17. flat rate: $1.25; per mile: $.20

18. flat rate: $1.95; per mile: $.10

Objective 3

19. 80 oz of 20%; 40 oz of 50%

20. 40 oz of 40%; 120 oz of 80%

21. $\dfrac{10}{9}$ L

22. 2.5 L

23. 15 LB

24. 80 kg of $12; 40 kg of $15

25. 60 kg of $10; 20 kg of $18

26. 20 lb of $4; 40 lb of $7

Objective 4

27. 32 mph, 64 mph

28. 10 mph

29. 5.5 hr

30. $1\dfrac{1}{8}$ mi

31. 20 mph

32. 565 mph

33. 5 mph

34. 225 mph

35. $1\dfrac{1}{2}$ mph

36. 60 kph

Objective 5

37. 9, 10, 12

38. 28, 31, 40

39. 50°, 70°, 60°

40. 40°, 65°, 75°

41. 11 in., 21 in., 28 in.

42. 7 fives, 12 tens, 32 twenties

43. 12 tens, 25 twenties, 13 fifties

44. $40,000 at 5%; $10,000 at 6%; $30,000 at 7%

45. $20,000 at 8%; $50,000 at 10%; $5000 at 11%

46. 15 lb of $8; 12 lb of $10; 23 lb of $15

5.3   Mixed Exercises

47. side of square: 8 cm;
side of triangle: 13 cm

48. dark clay: $5/kg;
light clay: $1/kg

49. $4000 at 6%;
$8000 at 3%

50. 3300 tickets

51. 21 cm, 28 cm, 32 cm

52. 10 gal of 15%;
20 gal of 30%

53. car: 75 kph;
truck: 65 kph

54. 20 lb of $4; 30 lb of $6;
50 lb of $10

55. current: 3 mph;
boat: 18 mph

56. 5 quarters

5.4 Solving Systems of Linear Equations by Matrix Methods

Objective 1

1. 2×2

2. 3×2

3. 2×3

4. 4×2

5. 4×3

6. 3×4

Objective 2

7. $\begin{bmatrix} 3 & -4 & | & 7 \\ 2 & 1 & | & 12 \end{bmatrix}$

8. $\begin{bmatrix} 2 & -3 & | & 12 \\ 7 & 3 & | & 15 \end{bmatrix}$

9. $\begin{bmatrix} \frac{1}{3} & -\frac{1}{2} & | & 7 \\ \frac{5}{3} & \frac{1}{2} & | & 8 \end{bmatrix}$

10. $\begin{bmatrix} \frac{1}{2} & \frac{1}{2} & | & -16 \\ -3 & 1 & | & 2 \end{bmatrix}$

11. $\begin{bmatrix} 3 & -1 & | & 4 \\ -5 & 1 & | & -2 \end{bmatrix}$

12. $\begin{bmatrix} 2 & 1 & | & -3 \\ 1 & -4 & | & -5 \end{bmatrix}$

13. $\begin{bmatrix} -2 & 3 & -5 & | & 7 \\ 6 & 2 & -4 & | & 12 \\ 5 & -2 & 1 & | & -1 \end{bmatrix}$

14. $\begin{bmatrix} 1 & 1 & 1 & | & 10 \\ 2 & 1 & -3 & | & 11 \\ 1 & 0 & 2 & | & -2 \end{bmatrix}$

Objective 3

15. $\{(2, -3)\}$

16. $\{(3, 2)\}$

17. $\{(3, -2)\}$

18. $\{(0, -2)\}$

19. $\{(0, 3)\}$

20. $\{(1, -1)\}$

21. $\{(3, 1)\}$

22. $\{(-5, -3)\}$

23. $\left\{ \left(3, \frac{3}{2} \right) \right\}$

Objective 4

24. $\{(2, -1, -3)\}$

25. $\{(2, -1, 4)\}$

26. $\{(1, 3, -2)\}$

27. $\{(3, -1, 4)\}$

28. $\{(1, 2, 4)\}$

29. $\{(-2, 1, 2)\}$

Objective 5

30. $\varnothing$

31. $\varnothing$

32. $\{(x, y) | x - 2y = 3\}$

33. $\{(x, y) | 2x + y = 10\}$

34. $\varnothing$

35. $\varnothing$

36. $\{(1, 0, 0)\}$

37. $\{(x, y, z) | x + 2y - z = 6\}$

5.4 Mixed Exercises

38. 2×3

39. 4×3

40. $\begin{bmatrix} 2 & -5 & | & 2 \\ 3 & 4 & | & -1 \end{bmatrix}$

41. $\begin{bmatrix} 2 & 1 & | & 7 \\ 1 & -4 & | & 0 \end{bmatrix}$

42. $\varnothing$

43. $\{(-2, -3)\}$

44. $\{(4, -2)\}$

45. $\{(x, y) \mid 2x - y = 2\}$

46. $\varnothing$

47. $\{(1, -1, -1)\}$

48. $\{(-4, 3, 2)\}$

Chapter 6

EXPONENTS AND POLYNOMIALS

6.1 Integer Exponents and Scientific Notation

Objective 1

1. 3^6 or 729

2. 2^8 or 256

3. x^8

4. x^{19}

5. y^9

6. x^6

7. y^{11}

8. $6x^9$

9. $12p^{11}$

10. $-12r^7$

11. $-12z^5$

12. $3x^{26}$

13. $63t^{17}$

14. $-18y^{19}$

15. $-28p^{18}$

Objective 2

16. 1

17. -1

18. 16

19. $-38k^2$

20. $-10p^2r^2$

21. $96x^2$

22. $\dfrac{1}{7^2}$ or $\dfrac{1}{49}$

23. $\dfrac{1}{10^5}$ or $\dfrac{1}{100,000}$

24. $\dfrac{3}{x^3}$

25. $-\dfrac{6}{y^2}$

26. w^5

27. $\dfrac{1}{p^3}$

28. 1

29. 3

30. $\dfrac{12}{n^2}$

31. $-80g^5$

Objective 3

32. 1

33. 8^2 or 64

34. $\dfrac{1}{8}$

35. $\dfrac{1}{x^7}$

36. r^7

37. $\dfrac{1}{p}$

38. $\dfrac{1}{5^{15}}$

39. x^5

40. 4^7

41. 5

42. $\dfrac{1}{y^9}$

43. 3^2 or 9

Objective 4

44. 3^8 or 6561

45. x^{16}

46. $\dfrac{3^3}{4^3}$ or $\dfrac{27}{64}$

47. $36x^2$

48. $64y^{12}$

49. $81x^8$

50. 1

51. a^{12}

52. $\dfrac{1}{3^6}$ or $\dfrac{1}{729}$

53. $\dfrac{1}{3^2 z^8}$ or $\dfrac{1}{9z^8}$

54. 9

55. $-\dfrac{x^{12}}{3^3}$ or $-\dfrac{x^{12}}{27}$

56. $\dfrac{6^4}{a^{12}}$ or $\dfrac{1296}{a^{12}}$

57. $\dfrac{1}{3^8 m^6}$ or $\dfrac{1}{6561 m^6}$

58. $x^5 y^3$

Objective 5

59. $\dfrac{1}{4^5}$

60. $\dfrac{1}{6^6}$

61. $\dfrac{1}{7^{10}}$

62. $18y^2$

63. $\dfrac{1}{9^8}$

64. $\dfrac{1}{9^8}$

65. $12w^3$

66. $\dfrac{9}{a^4 b^8}$

67. $\dfrac{25 j^8}{k^4}$

68. $\dfrac{x^{10}}{144 y^8}$

69. $\dfrac{b}{18}$

70. $\dfrac{x^2 y}{72}$

71. $\dfrac{r^8}{8}$

72. $\dfrac{1}{63x^8}$

73. $\dfrac{49}{16}$

74. $\dfrac{6}{5}$

75. $\dfrac{225 q^4}{64 p^6}$

76. $\dfrac{250 k^6}{27}$

Objective 6

77. 3.4×10^2

78. 3.4×10^{-3}

79. 1.4×10^1

80. 2.7×10^{-2}

81. 3.82×10^{-5}

82. 1.68×10^5

83. 1.68×10^{-1}

84. 4.72×10^{-2}

85. 8.3632×10^4

86. 9.3×10^7

87. 3.75×10^{-7}

88. 3.75×10^1

89. 342,000

90. .00082

91. 2220

92. .715

93. 4.169

94. 58,300

95. .0000624

96. 9,300,000

97. 2.0×10^4 or 20,000

98. 3.0×10^{-3} or .003

99. 2.0×10^7 or 20,000,000

100. 3.0×10^8 or 300,000,000

101. 5×10^{-7} or .0000005

102. 1.0×10^6 or 1,000,000

103. 350

104. 1×10^4 or 10,000

6.1   Mixed Exercises

105. $\dfrac{1}{4}$

106. 10^3 or 1000

107. 1

108. p^6

109. 4^9 or $262,144$

110. $\dfrac{1}{7^2}$ or $\dfrac{1}{49}$

111. $\dfrac{1}{m^7}$

112. t^4

113. $-\dfrac{20}{x^8}$

114. $\dfrac{3}{5}$

115. $-2^5 y^{20}$ or $-32y^{20}$

116. $\dfrac{1}{14^2}$ or $\dfrac{1}{196}$

117. $\dfrac{m^8}{p^6}$

118. $\dfrac{60}{z^2}$

119. $3^4 x^8$ or $81x^8$

120. $\dfrac{1}{125y^{18}}$

121. $\dfrac{8w^7}{9y^5}$

122. $\dfrac{4^2}{3^2}$ or $\dfrac{16}{9}$

123. $\dfrac{2^3}{9^2 r^3}$ or $\dfrac{8}{81r^3}$

124. $\dfrac{c}{d^8}$

125. $\dfrac{25s^{16}}{9r^8 t^8}$

126. 1.5×10^{-4} or $.00015$

127. 3.0×10^{-9} or $.000000003$

128. 3.0×10^{-1} or $.3$

129. $\dfrac{4}{3} \times 10^{-5}$ or $1.\overline{3} \times 10^{-5}$

6.2   Adding and Subtracting Polynomials

Objective 1

1. $256, 3$

2. $-12, 4$

3. $1, 4$

4. $-1, 6$

5. $-20, 5$

6. $67, 3$

7. descending powers

8. not

9. not

10. descending powers

11. binomial

12. trinomial

13. none of these

14. none of these

15. monomial

16. trinomial

17. binomial

18. binomial

19. monomial

20. binomial

Objective 2

21. 5

22. 1

23. 0

24. 3

25. 12

26. 4

27. 1

28. 10

29. 4

30. 2

Objective 3

31. $17x + 3$

32. $2x^2 + 8x - 9$

33. $-18z^2 + 9z - 3$

34. $-m^2 + 7m - 3$

35. $8a^3 - 2a^2 + 7a + 7$

36. $-2p^2 + 12p - 7$

37. $10x^2 + 3x + 4$

38. $2r^3 - 3r^2 - 2r$

39. $3x - 6$

40. $-7k + 13$

41. $-2n^2 - 4n + 4$

42. $8x^3 + 7x - 11$

43. $-y^4 - 2y^3 + 7y^2 - 3y$

44. $13r^3 + 12r - 11$

45. 2

46. $-x^4 - 2x^3 - 7x^2 + 14$

47. $13y$

6.2   Mixed Exercises

48. trinomial, 2

49. none of these, 6

50. binomial, 1

51. monomial, 0

52. $4x^2 + 11x - 9$

53. $-5m^3 + 2m^2 + 16m$

54. $x^5 + x^3 - 6x^2 + 2x$

55. $s^3 + s^2 + s + 1$

56. $-4n^3 + 7n^2 + 2n$

6.3   Polynomial Functions

Objective 1

1. 22

3. −5

5. 0

7. −17

9. −5

2. 72

4. −6

6. −7

8. −5

10. −7

Objective 2

11. (a) 631.8 thousand

(b) 649.4 thousand

12. (a) 15,056

(b) 17,801

13. (a) 49,459

(b) 52,927

Objective 3

14. (a) $8x - 4$

(b) $4x - 12$

15. (a) $5x + 3$

(b) $-11x + 1$

16. (a) $x^2 + 7x - 13$

(b) $3x^2 + x + 3$

17. (a) $3x^2 - 6x + 21$

(b) $9x^2 - 8x + 3$

Objective 4

18.

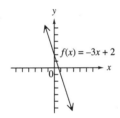

domain: $(-\infty, \infty)$; range: $(-\infty, \infty)$

20.

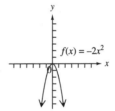

domain: $(-\infty, \infty)$; range: $(-\infty, 0]$

19.

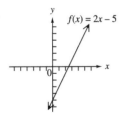

domain: $(-\infty, \infty)$; range: $(-\infty, \infty)$

21.

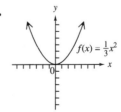

domain: $(-\infty, \infty)$; range: $[0, \infty)$

22.

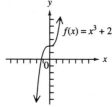

domain: $(-\infty, \infty)$; range: $(-\infty, \infty)$

23.

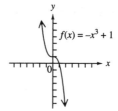

domain: $(-\infty, \infty)$; range: $(-\infty, \infty)$

6.3   Mixed Exercises

24. -197

25. -50

26. 7

27. (a) $-x^2 + 11x - 12$

 (b) $5x^2 + 7x + 2$

28.

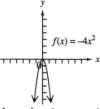

domain: $(-\infty, \infty)$;
range: $(-\infty, 0]$

29.

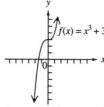

domain: $(-\infty, \infty)$;
range: $(-\infty, \infty)$

6.4   Multiplying Polynomials

Objective 1

1. $20a^2$

2. $40t^2$

3. $-60s^5$

4. $-10r^6$

5. $-28x^4$

6. $6k^6$

7. $72x^3y^4$

8. $-22x^4y^3$

9. $18y^3z^7$

10. $108r^5s^7$

Objective 2

11. $21n+28$

12. $3p^2-5p$

13. $-48y^2+56y$

14. $6x^4+8x^3+10x^2$

15. $-16z^6-32z^4$

16. $12m^3+17m^2-14m-15$

17. $3a^3-5a^2+5a-2$

18. $-5t^7-25t^6$

19. $24s^3+11s^2-39s-5$

20. $4x^4+2x^3+21x^2+10x+5$

Objective 3

21. $24r^2-8r-2$

22. $q^2+2q-35$

23. $14k^2-k-4$

24. n^2-n-20

25. $x^2+19x+84$

26. $x^2-3x-28$

27. $6z^2-7z+2$

28. $5y^2-14y-3$

29. $2m^2-11m+12$

30. $24t^2-10t-21$

Objective 4

31. $25a^2-1$

32. $4z^2-9$

33. $4x^2-25y^2$

34. $100r^2-s^2$

35. $81x^2-16y^2$

36. $25p^2-9q^2$

37. $64k^2-m^2$

38. $x^2-\dfrac{1}{4}$

39. $16t^6-1$

40. $4y^4-z^6$

Objective 5

41. $p^2-14p+49$

42. $r^2+10r+25$

43. $x^2+4xy+4y^2$

44. $x^2 - 14xy + 49y^2$

45. $36r^2 - 132r + 121$

46. $25m^2 + 20mn + 4n^2$

47. $4a^2 - 28ab + 49b^2$

48. $x^2 + x + \dfrac{1}{4}$

49. $4k^2 - 12km + 9m^2$

50. $9x^2 - 24xy + 16y^2$

Objective 6

51. $6x^2 - 3x$

52. $10x^2 - 14x$

53. $3x^2 + 4x - 4$

54. $3x^2 - 11x - 20$

55. $6x^3 - x^2 - 31x - 24$

56. $16x^3 + 28x^2 - 10x + 50$

6.4   Mixed Exercises

57. $-15x^5 y^7$

58. $s^2 - 4s + 3$

59. $24y^2 - 49yz + 15z^2$

60. $4p^2 + 4pq - 35q^2$

61. $49x^2 - 100$

62. $32q^6 - 12q^3 + 24q^2$

63. $9r^2 + 18r - 16$

64. $9s^4 - 1$

65. $2x^2 - xy - 3y^2$

66. $z^2 + 20z + 100$

67. $8x^2 - 19x - 15$

68. $6x^3 + x^2 + 17x + 9$

6.5 Dividing Polynomials

Objective 1

1. $4z^4$

2. $9y^5$

3. $6m^3n^4$

4. $7t^6 - 3t$

5. $10x^3 - 16 + 8x$

6. $2a^2 - 1 + \frac{1}{a}$

7. $5 - 3r$

8. $9 + \frac{3}{q} + \frac{9q^2}{5}$

9. $-4x^5 + 2x^4 + x^3 - 3x^2 + 1$

10. $\frac{7b}{5} - \frac{9}{5} + \frac{1}{b}$

11. $\frac{5}{3z^4} + \frac{1}{3z^3} + \frac{2}{z^2} + \frac{8}{3z}$

12. $\frac{12}{a} - \frac{6}{a^2} + \frac{14}{a^3}$

Objective 2

13. $m + 1$

14. $x + 3$

15. $z - 3$

16. $3x + 4$

17. $4s - 1 + \frac{-5}{s+3}$

18. $3r^2 - 5r + 6$

19. $4x - 3 + \frac{4x+5}{3x^2 - 2x + 5}$

20. $y^2 - 2 + \frac{5}{y^2 - 1}$

21. $p^3 + 3p^2 - p$

22. $2m^3 - m^2 + m + 1 + \frac{11}{4m^2 - 3}$

Objective 3

23. $5x - 1; 0$

24. $4x - 7; 0$

25. $2x - 5; -2$

26. $4x + 9; 5$

27. $9x^2 + 6x + 4; \frac{2}{3}$

6.5 Mixed Exercises

28. $\frac{2}{3} - \frac{8}{9x^2} + \frac{1}{x^3}$

29. $2q - 1 + \frac{4}{2q-1}$

30. $2k - 1 + \frac{6}{2k-1}$

31. $2s - 2 + \frac{5}{s}$

32. $x + 3$

33. $t + 2$

34. $9p^3 - 12p^2 - 2p + \frac{26}{3} - \frac{2}{3p}$

35. $x^2 + 2 + \dfrac{-3}{x^2 - 2}$

36. $n^3 + n + \dfrac{n - 2}{n^2 - 1}$

37. $-n^2 + 3n - \dfrac{4}{n}$

38. $6x - 1; 0$

39. $2x + 7; 6$

Chapter 7

FACTORING

7.1 Greatest Common Factors; Factoring by Grouping

Objective 1

1. $7(y+3)$

2. $8(4x-1)$

3. $15(x-y)$

4. $3(4a-b)$

5. cannot be factored

6. $2(13x+31y)$

7. $6a^2(2a+1)$

8. $16r^2(3+r^3)$

9. $(p-9)(2p+3)$

10. $(k-6)(4)$

11. $(2q+1)(-11)$

12. $(x+2)(x+2)$ or $(x+2)^2$

Objective 2

13. $(x+y)(1+6a)$

14. $(x+9)(x-y)$

15. $(3+a)(b+c)$

16. $(x+y)(4a+3b)$

17. $(x-1)(y-1)$

18. $(x-4)(1+2y^2)$

19. $(2-3p)(4-3p^3)$

20. $(x^3-3)(y^2+1)$

21. $(x+2)(y-3)$

22. $(x-y)(a+b)$

7.1 Mixed Exercises

23. $s^3(s^2+4s^3+8)$

24. $2(4m^5+3m^2-6)$

25. $5x^2y^2(x+5y)$

26. $7x^2y(2xy+1-3x^3y^2)$

27. cannot be factored

28. $a^2b^3(6+25a^2b^2)$

29. $(s-t)(5r-2q)$

30. $(7w+3x)(2w-5x)$

31. $(x^2+y^2)(3x+4y)$

32. $(4x^2-y)(4x-y^2)$

7.2   Factoring Trinomials

Objective 1

1. $(z+3)(z+2)$

2. $(y+3)(y-1)$

3. $(x+7)(x-4)$

4. $(m+3)(m-4)$

5. $(t+4)(t-5)$

6. $(x+6)(x-5)$

7. $(r+2)(r-8)$

8. $(q+5)(q+1)$

9. $(k+5)(k+4)$

10. $(n+3)(n+3)$

11. $(x+3y)(x+5y)$

12. $(a+5)(a-7)$

13. $(rs+7)(rs-3)$

14. $(xy-3)(xy-9)$

Objectives 2 and 3

15. $(3y+1)(y+4)$

16. $(4x+1)(x+1)$

17. $(3z-4)(z+2)$

18. $(3t-1)(2t+1)$

19. $(2r-3)(10r+1)$

20. $(5x+3)(x+2)$

21. $(5x+11)(4x-1)$

22. $(5q-4)(3q+1)$

23. $(4x-5)(2x+3)$

24. $(3p-5)(2p+3)$

25. $(3y+2z)(2y-3z)$

26. $(6x+y)(x-y)$

27. $(4k+p)(k+3p)$

28. $(3x-4y)(2x+5y)$

Objective 4

29. $(2x+2y+1)(x+y+3)$

30. $(3p-3q+7)(p-q+1)$

31. $(5a-2)(a-3)$

32. $(4p+15)(2p+13)$

33. $(19-4z)(7-2z)$

34. $(2r+2s+3)(5r+5s-8)$

35. $(8y^2-3)(5y^2+2)$

36. $(9x^2-5)(2x^2+1)$

37. $(4t^2+1)(t^2+17)$

38. $(3x^2y^2-5)(2x^2y^2+1)$

7.2   Mixed Exercises

39. $(x+3)(x-9)$

40. $(2a-5)(a-6)$

41. $(3m+n)(2m-5n)$

42. $(x+3y)(x-5y)$

43. $(2y^2+9)(2y^2-5)$

44. $(5x-2y)(5x+y)$

45. $(p+3)(p-7)$

46. $(a+8)(3a+1)$

47. $(3m-1)(4m+5)$

48. $(p+8)(p+4)$

7.3   Special Factoring

Objective 1

1. $(y+4)(y-4)$

2. $(3x+1)(3x-1)$

3. $(6z+11)(6z-11)$

4. cannot be factored

5. $(12x+5y)(12x-5y)$

6. $(a+10)(a-10)$

7. $(s^2+4)(s+2)(s-2)$

8. $(x+9)(x-9)$

9. $(x+y+5)(x+y-5)$

10. $(q+2r+3)(q-2r-3)$

11. $(5+x-y)(5-x+y)$

12. $-4rs$

Objective 2

13. $(x+2)^2$

14. $(z-5)^2$

15. $(x+4)^2$

16. $(y+11)^2$

17. $(x+12)^2$

18. $(k+6)^2$

19. $(4a-5b)^2$

20. $(4t+7)^2$

21. $(6r-5s)^2$

22. $(5x-2y)^2$

23. $(y+z+7)^2$

24. $(p-q-10)^2$

25. $(2x-y)^2$

26. $(m-n-6)^2$

Objective 3

27. $(x-y)(x^2+xy+y^2)$

28. $(2a-1)(4a^2+2a+1)$

29. $(2r-3s)(4r^2+6rs+9s^2)$

30. $(4x-y)(16x^2+4xy+y^2)$

31. $(6m-5p^2)(36m^2+30mp^2+25p^4)$

32. $(2a-5b)(4a^2+10ab+25b^2)$

33. $(r+s-1)(r^2+2rs+s^2+r+s+1)$

34. $2n(3m^2+n^2)$

35. $3x^2-3x+1$

36. $8(3x-y)(9x^2+3xy+y^2)$

Objective 4

37. $(x+y)(x^2-xy+y^2)$

38. $(z+2)(z^2-2z+4)$

39. $(3r+2s)(9r^2-6rs+4s^2)$

40. $8(a+2b)(a^2-2ab+4b^2)$

41. $(5p+q)(25p^2-5pq+q^2)$

42. $(4x+7y)(16x^2-28xy+49y^2)$

43. $(1+y+z)(1-y-z+y^2+2yz+z^2)$

44. $2x(x^2+3y^2)$

45. $(2a-1)(a^2-a+1)$

46. $2(t+1)(t^2+2t+4)$

7.3   Mixed Exercises

47. $(x-7)^2$

48. $(x^2+9)(x+3)(x-3)$

49. $(5m+2n)(25m^2-10mn+4n^2)$

50. $(4+p-q)(4-p+q)$

51. 0

52. $(y^2+1)(y^4-y^2+1)$

53. $(5m-4p)(25m^2+20mp+16p^2)$

54. $(2x+3y)^2$

55. $(y+z)(y-z)$

56. $(z-5y)(z^2+5yz+25y^2)$

7.4 Solving Equations by Factoring

Objective 1

1. $\left\{-\dfrac{3}{2}, 4\right\}$

2. $\{-4, 6\}$

3. $\left\{-\dfrac{3}{4}, 12\right\}$

4. $\{0, 3\}$

5. $\{-9, 8\}$

6. $\{-5, 4\}$

7. $\left\{-\dfrac{5}{2}, -\dfrac{2}{3}\right\}$

8. $\left\{\dfrac{5}{4}, -\dfrac{3}{2}\right\}$

9. $\left\{-\dfrac{5}{4}, \dfrac{1}{3}\right\}$

10. $\left\{-\dfrac{2}{5}, 2\right\}$

11. $\left\{-\dfrac{3}{2}, -\dfrac{1}{3}\right\}$

12. $\left\{0, -\dfrac{3}{5}\right\}$

13. $\left\{-3, \dfrac{3}{2}\right\}$

14. $\left\{-\dfrac{6}{5}, \dfrac{6}{5}\right\}$

Objective 2

15. 3 and 4

16. 0 and 12

17. 12 and 13

18. 12 and 14

19. 11 and 13

20. 16 and 18

21. 10 and −7

22. −8 and 2

23. 7 ft

24. 19 m

7.4 Mixed Exercises

25. $\left\{-\dfrac{3}{4}, 2\right\}$

26. $\left\{-\dfrac{5}{4}, \dfrac{5}{4}\right\}$

27. $\{-2, 2\}$

28. $\{3\}$

29. $\left\{-\dfrac{1}{3}, \dfrac{2}{5}\right\}$

30. $\left\{-\dfrac{7}{3}, 4\right\}$

31. 10 and −6

32. 10 ft

33. 16 ft by 25 ft

34. 18 ft by 30 ft

Chapter 8

RATIONAL EXPRESSIONS

8.1 Rational Expressions and Functions; Multiplying and Dividing

Objectives 1 and 2

1. 5

2. –5

3. –7

4. 0

5. None

6. $\dfrac{7}{4}$

7. $\dfrac{2}{3}$

8. 5, –5

9. 1, 2

10. 5

11. None

12. None

Objective 3

13. $\dfrac{n^2}{4}$

14. $\dfrac{7p^5}{3}$

15. $-\dfrac{x^2 y^3}{3}$

16. $\dfrac{m-7}{m-2}$

17. $\dfrac{1}{5}$

18. $\dfrac{3}{4}$

19. $\dfrac{6y+1}{3y+1}$

20. $\dfrac{x-3}{x+3}$

21. $\dfrac{11r^2}{6}$

22. $\dfrac{s+2}{s+4}$

23. $\dfrac{2z+3}{4z+3}$

24. 1

25. –1

26. $-(x+1)$ or $-x-1$

27. –1

28. $\dfrac{r^2 + rs + s^2}{r+s}$

29. $\dfrac{x+y}{x^2 + xy + y^2}$

30. $\dfrac{p-5}{5+p}$

Objective 4

31. $\dfrac{3x}{4}$

32. $\dfrac{3z}{2}$

33. $\dfrac{3}{8}$

34. $8y$

35. $2a^4$

36. $\dfrac{6}{m+3}$

37. $\dfrac{6}{5}$

38. 1

39. $\dfrac{3}{10}$

40. $\dfrac{2x}{3}$

41. $\dfrac{x+4}{x-4}$

42. $\dfrac{2z-3}{2z+3}$

Objective 5

43. $\dfrac{y}{3}$

44. $\dfrac{p-5}{4}$

45. $\dfrac{7x}{x^2+9}$

46. $\dfrac{5}{m^2+2m+3}$

47. $\dfrac{p^2+7p}{2p-1}$

48. $\dfrac{7}{n+8}$

49. –1

50. $\dfrac{5+r}{r^2+2r}$

51. $\dfrac{3x-6}{x^2+4}$

52. $\dfrac{z^2-9}{7z+7}$

53. No reciprocal for 0

54. $\dfrac{x^2+x+2}{x^2-3x+4}$

Objective 6

55. $\dfrac{1}{4}$

56. $\dfrac{25t^3}{9}$

57. $\dfrac{8s^2}{3}$

58. $\dfrac{2r^3}{3}$

59. 2

60. $\dfrac{3}{2}$

61. $\dfrac{2}{9}$

62. $\dfrac{4}{9}$

63. $(a+4)(a-3)$ or a^2+a-12

64. $\dfrac{18}{(m-1)(m+2)}$ or $\dfrac{18}{m^2+m-2}$

65. $\dfrac{2(z+4)}{z-3}$

66. $-\dfrac{y+8}{y-8}$ or $\dfrac{y+8}{8-y}$

67. $\dfrac{x+2}{x+3}$

68. 1

69. $\dfrac{s+3}{s+4}$

70. $\dfrac{a-3}{2a-3}$

8.1 Mixed Exercises

71. $\dfrac{q}{5r^2}$

72. $-\dfrac{3}{2}$

73. $-z$

74. r

75. $-\dfrac{7}{8}$

76. -3

77. $(t+1)(t+2)$ or t^2+3t+2

78. $-x$

79. $\dfrac{p-6}{p+3}$

80. $\dfrac{(z+4)^3}{z(z-16)}$

8.2   Adding and Subtracting Rational Expressions

Objective 1

1. $\dfrac{11}{x}$

2. $-\dfrac{3}{y^2}$

3. $\dfrac{18}{5t}$

4. $\dfrac{c-4}{5a}$

5. $\dfrac{4n-7}{m+3}$

6. $z+y$

7. $\dfrac{1}{r-s}$

8. 1

9. $\dfrac{1}{x-5}$

10. $\dfrac{1}{k+5}$

11. $\dfrac{1}{q-7}$

12. $-\dfrac{1}{a+b}$

Objective 2

13. $30m$

14. $150z$

15. $75x^2y$

16. $t(t-1)$

17. $24(s+3)$

18. $5a(a+2)$

19. $(q-6)(q+6)^2$

20. $r-p \text{ or } p-r$

21. $r(r+4)(r+1)$

22. $n(3+n)(3-n)$

23. $(p-4)(p+4)^2$

24. $(2z-1)(z+4)(z-3)$

Objective 3

25. $\dfrac{4y+35}{7y}$

26. $\dfrac{18+3x}{2x}$

27. $\dfrac{3z-5}{5z}$

28. $\dfrac{a}{3}$

29. $\dfrac{2+m}{2}$

30. $\dfrac{6-2s}{s^2}$

31. $\dfrac{18t+20}{15t^2}$

32. $2r+2$

33. $\dfrac{3}{(x+1)(x-1)(x+2)}$

34. $\dfrac{3y+1}{y^2-16}$

35. $\dfrac{3}{a-2} \text{ or } -\dfrac{3}{2-a}$

36. $-\dfrac{1}{m-3} \text{ or } \dfrac{1}{3-m}$

37. $\dfrac{8r}{r+2s}$

38. $\dfrac{2(a^2+3ab+4b^2)}{(b+a)^2(3b+a)}$

8.2   Mixed Exercises

39. $m - 2$ or $2 - m$

40. $3(x + 2)(x + 3)$

41. $\dfrac{1}{x + 2}$

42. $\dfrac{3}{r - 6}$ or $-\dfrac{3}{6 - r}$

43. $\dfrac{8ab + 21}{6a^2 b}$

44. $\dfrac{2q^2 + 4q + 4}{q(q + 2)}$

45. $\dfrac{4k + 24}{(k - 2)(k + 2)}$

46. $\dfrac{8m - 4n}{m^2 - n^2}$

47. $\dfrac{2z^2 - z + 1}{(z - 1)(z + 1)^2}$

48. $\dfrac{3y - 5}{2(y + 2)(y - 2)}$

8.3   Complex Fractions

Objective 1

1. $\dfrac{y}{x}$

2. $\dfrac{3}{20}$

3. $\dfrac{2}{k}$

4. $\dfrac{(z+t)^2}{tz}$

5. $\dfrac{r}{s}$

6. $\dfrac{(m+1)^2}{m-1}$

7. $\dfrac{q(q^2+q+1)}{(q+1)^2(q-1)}$

8. $\dfrac{a-b}{4}$

9. $\dfrac{1}{rs}$

Objective 2

10. $\dfrac{k}{3}$

11. $\dfrac{m}{2}$

12. $\dfrac{3}{x}$

13. $\dfrac{b}{a}$

14. $\dfrac{2s}{5}$

15. $\dfrac{p^2+1}{3-p^2}$

16. $\dfrac{t^2-2}{t^2+4}$

17. $\dfrac{x+1}{1-x}$

18. $\dfrac{r-1}{r+1}$

Objective 3

19. n

20. $\dfrac{a^2+1}{1-a^2}$

21. $\dfrac{(z-y)^2}{z(z+y)}$

22. $\dfrac{-t+1}{t+1}$

23. $\dfrac{7(5k-m)}{4}$ or $\dfrac{35k-7m}{4}$

24. $\dfrac{1-r}{r(1+r)}$

Objective 4

25. $\dfrac{1}{5x+1}$

26. $\dfrac{x^2+xy+x+y}{x}$

27. $\dfrac{y^2+x}{xy^2}$

28. $\dfrac{2z^3+xy^2z^3}{x}$

29. $\dfrac{y^3}{4y^3z^2+z^2}$

30. $\dfrac{x^2y^2}{y^2-x^2}$

31. $\dfrac{2y^3}{x^2y^3+3x^2}$

32. $\dfrac{r}{s}$

33. $\dfrac{1}{xy-1}$

34. $\dfrac{m^2n^2}{(m+n)^2(n^2-m^2)}$

8.3   Mixed Exercises

35. $\dfrac{2x-1}{-x-1}$ or $-\dfrac{2x-1}{x+1}$

36. $\dfrac{1}{p}$

37. $\dfrac{n}{m(n+1)}$

38. $\dfrac{n(3n-2)}{3n-10}$

39. $\dfrac{m-2}{8}$

40. $\dfrac{s(s-1)}{2}$

41. $\dfrac{y-x}{x+y}$

42. $\dfrac{x^4}{4x^4+1}$

43. $1-x$

44. $\dfrac{y-x}{x^2y^2}$

8.4   Equations Involving Rational Expressions

Objective 1

1. (a) $0, -1$

   (b) $\{x \mid x \neq 0, -1\}$

2. (a) $\dfrac{5}{2}, -\dfrac{1}{3}$

   (b) $\left\{x \mid x \neq \dfrac{5}{2}, -\dfrac{1}{3}\right\}$

3. (a) $7, -8$

   (b) $\{x \mid x \neq 7, -8\}$

4. (a) -1

   (b) $\{x \mid x \neq -1\}$

5. (a) $0, 2$

   (b) $\{x \mid x \neq 0, 2\}$

6. (a) $-3, 0, 1$

   (b) $\{x \mid x \neq -3, 0, 1\}$

Objective 2

7. $\left\{\dfrac{11}{4}\right\}$

8. $\{-7\}$

9. $\{1\}$

10. $\{5\}$

11. $\{-6\}$

12. $\{-8\}$

13. $\{2\}$

14. $\{4, 6\}$

15. $\{3\}$

16. $\left\{-\dfrac{4}{3}, 1\right\}$

Objective 3

17.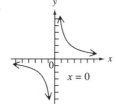

18.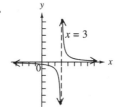

19.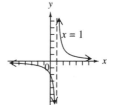

20.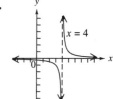

8.4   Mixed Exercises

21. $\left\{\dfrac{1}{3}\right\}$

22. $\left\{\dfrac{64}{3}\right\}$

23. $\{-3\}$

24. $\{-2\}$

25. $\varnothing$

26. $\{-6\}$

27. $\left\{\dfrac{4}{3}\right\}$

28. $\{2, 9\}$

29. $\left\{-6, \dfrac{1}{2}\right\}$

30. $\varnothing$

8.5   Applications of Rational Expressions

Objective 1

1. $b = \dfrac{5}{9}$

2. $R = \dfrac{1}{3}$

3. $m = \dfrac{75}{8}$

4. $L = 20$

5. $R_2 = 20$

6. $d_i = \dfrac{50}{3}$

7. $t = 2$

8. $r = \dfrac{1}{20}$ or .05 or 5%

9. $B = 7$

10. $h = 12$

Objective 2

11. $R_2 = \dfrac{RR_1}{R_1 - R}$

12. $d_0 = \dfrac{fd_i}{d_i - f}$

13. $T_2 = \dfrac{T_1 V_2 P_2}{V_1 P_1}$

14. $m_2 = \dfrac{Fd^2}{Gm_1}$

15. $a_n = \dfrac{2s_n}{n} - a_1$

16. $b_1 = \dfrac{2A}{h} - b_2$

17. $V_1 = \dfrac{T_1 V_2 P_2}{P_1 T_2}$

18. $R_1 = \dfrac{AR_2}{R_2 - A}$

19. $E = \dfrac{nE - Inr}{I}$

20. $r = \dfrac{eR}{E - e}$

Objective 3

21. 20 questions

22. 1750 crimes

23. $\dfrac{2}{9}$ job/hr

24. $38.40

25. $1125

26. 5 hr

27. 8 gal

28. $3.90

29. 50,000

30. $1000

Objective 4

31. 30 mi

32. 2 mph

33. 16 mi

34. 8 mph

35. 60 mph

36. Pauline: 60 mph;
Pete: 40 mph

40. Ted: 4 mph;
Olivia: 8 mph

37. 60 mi

38. 12 km/hr

39. 120 mi

Objective 5

41. $\dfrac{12}{7}$ or $1\dfrac{5}{7}$ hr

42. $\dfrac{9}{4}$ or $2\dfrac{1}{4}$ hr

43. $\dfrac{88}{19}$ or $4\dfrac{12}{19}$ hr

44. $\dfrac{143}{24}$ or $5\dfrac{23}{24}$ min

45. 60 hr

46. 60 hr

47. 24 hr

48. $\dfrac{15}{8}$ or $1\dfrac{7}{8}$ hr

49. $\dfrac{4}{5}$ hr or 48 min

50. $\dfrac{30}{11}$ or $2\dfrac{8}{11}$ hr

8.5   Mixed Exercises

51. $C = 37$

52. $b = 12$

53. $z = -\dfrac{15}{2}$

54. $p = .8$

55. $F = \dfrac{9C}{5} + 32$

56. $x = \dfrac{48yz}{35z + 120y}$

57. $r = \dfrac{56p}{8 + p}$

58. $B = \dfrac{2A - hb}{h}$

59. $v_s = \dfrac{Fv - fv - fv_0}{F}$

60. $v_0 = \dfrac{Fv - Fv_s - fv}{f}$

61. $\dfrac{20}{9}$ or $2\dfrac{2}{9}$ hr

62. 4 mph

63. 20 baskets

64. 36 hr

65. 14 mph

66. $3.84

67. 4 mph

68. $188.10

69. 275 mph

70. 2 hr

Chapter 9

ROOTS, RADICALS, AND ROOT FUNCTIONS

9.1 Radical Expressions and Graphs

Objective 1

1. 3

2. 6

3. −2

4. 2

5. not a real number

6. 3

7. −5

8. 2

9. 5

10. −1

11. 15

12. −4

13. 2

14. 3

15. 4

16. −7

17. 9

18. 8

19. −11

20. 11

Objective 2

21. 3

22. 4

23. −3

24. 1

25. −3

26. $\left| x^3 \right|$

27. $\left| x^3 \right|$

28. y^2

29. $-a^3$

30. −2

31. r^2

32. $-c^2$

33. 5

34. c^9

35. −2

36. a^2

37. 2

38. −2

39. −1

40. 7

41. −12

42. $-\dfrac{7}{4}$

43. $-\dfrac{11}{5}$

44. 5

45. $\left| t \right|$

46. s^8

47. p^{10}

48. a^{18}

Objective 3

49.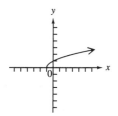

domain: $[-1, \infty)$
range: $[0, \infty)$

50.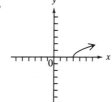

domain: $[3, \infty)$
range: $[0, \infty)$

51.

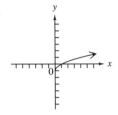

domain: $[0, \infty)$
range: $[-1, \infty)$

53.

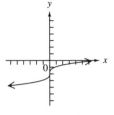

domain: $(-\infty, \infty)$
range: $(-\infty, \infty)$

52.

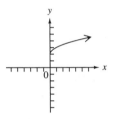

domain: $[0, \infty)$
range: $[2, \infty)$

54.

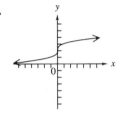

domain: $(-\infty, \infty)$
range: $(-\infty, \infty)$

Objective 4

55. $|x|$

56. $-|x|$

57. x

58. $-x$

59. x^2

60. $-x^2$

61. $-x^4$

62. x^3

Objective 5

63. 9.327

64. 8.718

65. -3.936

66. -3.271

67. 2.546

68. 3.027

69. 10.050

70. -12.610

71. 2.897

72. 6.407

73. -17.607

74. 2.418

75. 8.707

76. 29.496

77. 30.496

78. -3.377

79. -19.235

80. -3.973

81. 11.091

82. 3.037

9.1   Mixed Exercises

83. 14

84. 8

85. -15

86. 19.596

87. -18.248

88. 5.180

89. k^2

90. $|d|$

91. $-p^2$

92. 3.503

93. not a real number

94. 2.590

9.2 Rational Exponents

Objective 1

1. 3

2. 5

3. 2

4. 7

5. −3

6. not a real number

7. 10

8. −2

9. 5

10. 2

11. −4

12. 2

13. 6

14. 3

15. 15

16. 8

17. 9

18. 5

19. 6

20. −5

Objective 2

21. 9

22. 125

23. 243

24. $\dfrac{1}{9}$

25. $\dfrac{1}{125}$

26. $\dfrac{1}{9}$

27. 100

28. $-\dfrac{1}{25}$

29. 8

30. $\dfrac{1}{36}$

31. $\dfrac{1}{27}$

32. $\dfrac{1}{3}$

33. 8

34. $-\dfrac{1}{4}$

35. $\dfrac{1}{4}$

36. −125

37. $\dfrac{1}{64}$

38. 7776

39. 243

40. $\dfrac{1}{2}$

Objective 3

41. x^6

42. $\sqrt[3]{2t^2}$

43. $\sqrt[6]{x^5}$

44. 7^2 or 49

45. $x\sqrt{2x}$

46. $\sqrt[8]{y^7}$

47. $\sqrt[4]{5^3}$

48. $\sqrt[12]{2^7}$

49. $\sqrt[6]{x^5}$

50. $p\sqrt[5]{p^2}$

51. $\sqrt[6]{r}$

52. $\sqrt[5]{3t^2}$

53. $\sqrt[6]{k^5}$

54. $\sqrt[15]{p^{11}}$

55. $r^2\sqrt[3]{r}$

56. $x\sqrt[12]{x}$

57. $\sqrt[12]{x^4 y^3 z^6}$

58. $\sqrt[4]{20}$

Objective 4

59. 13^2 or 169

60. $\dfrac{1}{5^{1/7}}$

61. 8

62. 7^4 or 2401

63. 3^2 or 9

64. 5^3 or 125

65. $r^{11/12}$

66. y

67. $a^{5/6}$

68. a^4

69. 5^8 or 390,625

70. a

71. $x^{19/12}$

72. 8

73. $a^{2/15}$

74. $\dfrac{c^4}{x^2}$

75. $\dfrac{y^{10/3}}{x^{14/5}}$

76. $\dfrac{y}{x^{11/3}}$

9.2 Mixed Exercises

77. -3

78. 2

79. $-\dfrac{8}{125}$

80. $\dfrac{64}{27}$

81. -100

82. $-\dfrac{1}{8}$

83. $\dfrac{1}{125}$

84. 100

85. 1

86. x^2

87. $x\sqrt[9]{x}$

88. $\sqrt[15]{x}$

9.3 Simplifying Radical Expressions

Objective 1

1. $\sqrt{33}$

2. $\sqrt{70}$

3. $\sqrt{2t}$

4. $\sqrt{42xt}$

5. $\sqrt{\dfrac{91}{cw}}$

6. $\sqrt{\dfrac{33}{rp}}$

7. $\sqrt[4]{42}$

8. $\sqrt[6]{20t^5}$

9. $\sqrt[5]{24r^4t^4}$

10. $\sqrt[3]{21}$

11. $\sqrt{35}$

12. $\sqrt{165}$

13. $\sqrt{6r}$

14. $\sqrt[3]{5x}$

15. $\sqrt[3]{35xy}$

16. $\sqrt[4]{120}$

17. $\sqrt[7]{48a^3t^6}$

18. $\sqrt[4]{4x}$

Objective 2

19. $\dfrac{5}{4}$

20. $\dfrac{\sqrt{5}}{6}$

21. $\dfrac{\sqrt{x}}{9}$

22. $\dfrac{t^4\sqrt{t}}{5}$

23. $-\dfrac{3}{2}$

24. $\dfrac{\sqrt[3]{45}}{3}$

25. $\dfrac{7}{10}$

26. $\dfrac{\sqrt{6}}{7}$

27. $\dfrac{5}{r^5}$

28. $-\dfrac{a^2}{5}$

29. $\dfrac{w}{6}$

30. $\dfrac{\sqrt[3]{\ell^2}}{3}$

31. $\dfrac{\sqrt{r}}{11}$

32. $\dfrac{\sqrt[4]{p}}{2}$

33. $-\dfrac{7}{5}$

34. $\dfrac{z^2}{6}$

35. $\dfrac{\sqrt[5]{7x}}{2}$

36. $\dfrac{\sqrt{15}}{13}$

Objective 3

37. $3\sqrt{3}$

38. $3\sqrt{7}$

39. $5\sqrt{3}$

40. $10\sqrt{2}$

41. $3\sqrt{15}$

42. $2\sqrt[3]{3}$

43. $2\sqrt[4]{2}$

44. $6t^2\sqrt{t}$

45. $5r^2\sqrt{2rx}$

46. $2y^2\sqrt[3]{3}$

47. $3x^3\sqrt[3]{2x^2}$

48. $2x^3c^2\sqrt[3]{10c}$

49. $3a\sqrt[3]{4a^2}$

50. $7r^4\sqrt{r}$

51. $3bc^2\sqrt[3]{10bc^2}$

Objective 4

52. $x\sqrt[3]{x}$

53. $x\sqrt{x}$

54. $\sqrt[6]{5}$

55. $x^2\sqrt{x}$

56. $\sqrt[3]{9}$

57. $\sqrt[3]{x^2}$

58. $z\sqrt[4]{z}$

59. $\sqrt[4]{x^3y^2}$

60. $\sqrt[3]{z^2y}$

61. $\sqrt[4]{9x^2y^3}$

62. $\sqrt[6]{5400}$

63. $\sqrt[6]{63}$

Objective 5

64. 10

65. 26

66. $2\sqrt{2}$

67. $2\sqrt{6}$

68. $\sqrt{13}$

69. $6\sqrt{2}$

70. $4\sqrt{5}$

71. $4\sqrt{6}$

72. $\sqrt{39}$

73. $3\sqrt{5}$

Objective 6

74. $2\sqrt{5}$

75. $2\sqrt{13}$

76. $\sqrt{10}$

77. $\sqrt{61}$

78. $\sqrt{34}$

79. $5\sqrt{2}$

80. $\sqrt{58}$

81. $\sqrt{181}$

82. $\sqrt{37}$

83. $4\sqrt{3}$

84. $\sqrt{x^2+4y^2}$

85. $\sqrt{a^2-2ab+5b^2}$

9.3   Mixed Exercises

86. $\sqrt{30ab}$

87. $2x^3y^2\sqrt{y}$

88. $\sqrt[5]{8w^4}$

89. $3t^2\sqrt[3]{2t}$

90. 5

91. $x^3\sqrt{3xy}$

92. $\dfrac{3}{2}$

93. $\dfrac{a^2}{25}$

94. $\dfrac{b^4q^2\sqrt{b}}{2}$

95. $2xy^3z^5\sqrt{2xz}$

96. $v^3\sqrt[4]{t^3}$

97. $2\sqrt[10]{972}$

98. 13

99. $\sqrt{105}$

100. $2\sqrt{17}$

101. 5

9.4   Adding and Subtracting Radical Expressions

Objective 1

1. $5\sqrt{7}$

2. $13\sqrt{13}$

3. $27\sqrt{3}$

4. $49\sqrt{2}$

5. $22\sqrt[3]{3}$

6. $5\sqrt[4]{2}$

7. $3\sqrt{13}$

8. $-15\sqrt{6}$

9. $9\sqrt{3y}$

10. $7\sqrt{2}$

11. $9\sqrt[3]{3}$

12. $6\sqrt[3]{5}$

13. $12\sqrt{x}$

14. $-4\sqrt[3]{3}$

15. $12\sqrt{2}$ in., 18 in.^{2}

16. $11\sqrt{5}$ cm

17. $10\sqrt{3}$ ft, 18 ft^2

18. $35\sqrt{15}$ cm^2

19. $9\sqrt{10}$ m^2

9.4   Mixed Exercises

20. $\sqrt{10}$

21. $-\sqrt{6}$

22. $-3\sqrt[3]{2}$

23. $24\sqrt[3]{5}$

24. $5\sqrt[3]{2r}$

25. $13\sqrt{2z}$

26. $13\sqrt{2z}$

27. $28z\sqrt[3]{2}$

28. $3\sqrt{3}$ cm, 135 cm^2

29. $2\sqrt{35}$ in.

9.5   Multiplying and Dividing Radical Expressions

Objective 1

1. $6 + 3\sqrt{7} + 2\sqrt{2} + \sqrt{14}$

2. $2\sqrt{15} - \sqrt{110} + 3\sqrt{2} - \sqrt{33}$

3. 23

4. $\sqrt{10} - 4\sqrt{5} + 2\sqrt{3} - 4\sqrt{6}$

5. -14

6. $52 + 30\sqrt{3}$

7. $-16 + 13\sqrt{2}$

8. $14 - 4\sqrt{6}$

9. 4

10. $x - y^2$

11. $6x - 13\sqrt{x} + 6$

12. $4 - \sqrt[3]{25}$

Objective 2

13. $\dfrac{6\sqrt{5}}{5}$

14. $3\sqrt{5}$

15. $\dfrac{\sqrt{22}}{11}$

16. $\dfrac{\sqrt{30}}{2}$

17. $\dfrac{\sqrt{2}}{2}$

18. $\dfrac{3\sqrt{2}}{4}$

19. $\dfrac{7\sqrt{3}}{15}$

20. $\dfrac{-14\sqrt{3}}{9}$

21. $\dfrac{6\sqrt{2}}{5}$

22. $\dfrac{2\sqrt{6}}{3}$

23. $\dfrac{7\sqrt{3}}{12}$

24. $\dfrac{\sqrt{15}}{40}$

25. $\dfrac{\sqrt{5}}{2}$

26. $\dfrac{\sqrt{10}}{2}$

27. $\dfrac{3\sqrt{2}}{14}$

28. $\dfrac{3}{4}$

29. $\dfrac{6\sqrt{t}}{t}$

30. $\dfrac{5\sqrt{2r}}{r}$

31. $\dfrac{9x^2\sqrt{2t}}{t^3}$

32. $\dfrac{2\sqrt{2m}}{m}$

33. $\dfrac{5\sqrt{y}}{y}$

34. $\dfrac{\sqrt{10x}}{4}$

35. $\dfrac{2x\sqrt{3}}{3}$

36. $\dfrac{3\sqrt{14z}}{7}$

37. $\dfrac{2\sqrt{7y}}{7y}$

38. $\dfrac{\sqrt{10}}{4x}$

39. $\dfrac{2a\sqrt{15t}}{5t^2}$

40. $\dfrac{y\sqrt{21b}}{6b}$

41. $\dfrac{\sqrt{38}}{8}$

42. $\dfrac{\sqrt{5}}{5}$

43. $\dfrac{\sqrt[3]{3}}{3}$

44. $\dfrac{\sqrt[3]{42}}{9}$

45. $\dfrac{\sqrt[3]{10}}{5}$

46. $\dfrac{\sqrt[3]{xy^2}}{y}$

47. $\dfrac{\sqrt[3]{3}}{2}$

48. $\dfrac{\sqrt[3]{20x}}{2}$

49. $\dfrac{\sqrt[3]{42}}{6}$

50. $\dfrac{\sqrt[3]{100}}{5}$

51. $\dfrac{c\sqrt[3]{d}}{d}$

52. $\dfrac{t^2\sqrt[3]{x^2}}{x^3}$

53. $\dfrac{\sqrt[3]{28r}}{14}$

54. $\dfrac{\sqrt[3]{4mx^2}}{2x}$

Objective 3

55. $\dfrac{5\left(7+\sqrt{3}\right)}{46}$

56. $3\left(\sqrt{7}-3\right)$

57. $\dfrac{5\left(3+\sqrt{5}\right)}{4}$

58. $\dfrac{3\left(6-2\sqrt{2}\right)}{7}$ or $\dfrac{6\left(3-\sqrt{2}\right)}{7}$

59. $2\left(\sqrt{11}-\sqrt{2}\right)$

60. $-2\left(\sqrt{7}+\sqrt{5}\right)$

61. $\dfrac{-5\left(\sqrt{3}-\sqrt{11}\right)}{8}$

62. $\dfrac{4-\sqrt{5}}{11}$

63. $\dfrac{4+\sqrt{7}}{9}$

64. $-2\left(\sqrt{2}+\sqrt{3}\right)$

65. $\dfrac{7\left(\sqrt{3}+1\right)}{2}$

66. $-2\left(\sqrt{6}+\sqrt{3}\right)$

67. $\dfrac{-5\left(\sqrt{5}+\sqrt{3}\right)}{2}$

68. $\dfrac{\sqrt{3}\left(\sqrt{5}+\sqrt{2}\right)}{3}$ or $\dfrac{\sqrt{15}+\sqrt{6}}{3}$

69. $\dfrac{\sqrt{6}\left(\sqrt{13}-\sqrt{5}\right)}{8}$ or $\dfrac{\sqrt{78}-\sqrt{30}}{8}$

Objective 4

70. $4-\sqrt{2}$

71. $\dfrac{6-9\sqrt{3}}{4}$

72. $\dfrac{3+2\sqrt{15}}{4}$

73. $5-\sqrt{6}$

74. $\dfrac{1-\sqrt{2}}{2}$

75. $\dfrac{1+2\sqrt{3}}{5}$

76. $2+3\sqrt{3}$

77. $\dfrac{1-\sqrt{2}}{2}$

78. $2-\sqrt{3}$

79. $\dfrac{25+2\sqrt{5x}}{5}$

80. $3+\sqrt{5}$

81. $\dfrac{2-9\sqrt{2}}{3}$

9.5   Mixed Exercises

82. $3\sqrt{6}-12$

83. $\sqrt[3]{4}-1$

84. $\dfrac{-15\sqrt{2}}{4}$

85. $\dfrac{4\sqrt{3}}{3}$

86. $\dfrac{3\sqrt{2t}}{t}$

87. $\dfrac{-3\sqrt{3}-21+\sqrt{21}+7\sqrt{7}}{46}$

88. $\dfrac{6\sqrt{tx}-3x}{4t-x}$

89. $\dfrac{t-3\sqrt{t}}{t-9}$

90. $\dfrac{3\sqrt{3t}}{t^2}$

91. $\dfrac{\sqrt{30}}{10}$

92. $\dfrac{\sqrt{c}}{x}$

93. $\dfrac{\sqrt{15ab}}{3b}$

94. $\dfrac{x\sqrt{y}}{y^2}$

95. $\dfrac{a\sqrt[3]{ab^2}}{b}$

96. $\dfrac{\sqrt[3]{36s^2}}{6}$

97. $\dfrac{t^5\sqrt[3]{x^2}}{x}$

98. $\dfrac{\sqrt[3]{18xz^2}}{3z}$

99. $\dfrac{4-2\sqrt{2}}{3}$

100. $\dfrac{1-y\sqrt{2y}}{2}$

101. $\dfrac{6+\sqrt{5}}{5}$

102. $1+2\sqrt{2}$

103. $1+\sqrt{2x}$

104. $\dfrac{\sqrt{5}-1}{7}$

9.6 Solving Equations with Radicals

Objective 1

1. $\{25\}$
2. $\{2\}$
3. $\{6\}$
4. $\{8\}$
5. $\{10\}$
6. $\{0\}$
7. $\{15\}$
8. $\{1\}$
9. $\{10\}$

10. $\{11\}$
11. $\{6\}$
12. $\{16\}$
13. $\{41\}$
14. $\{4\}$
15. $\{100\}$
16. $\{15\}$
17. $\{7\}$
18. $\{21\}$

19. $\{7\}$
20. $\{20\}$
21. $\varnothing$
22. $\{16\}$
23. $\{23\}$
24. $\varnothing$
25. $\{15\}$
26. $\{5\}$
27. $\varnothing$

28. $\{13\}$
29. $\{2\}$
30. $\{1\}$
31. $\{22\}$
32. $\{11\}$
33. $\{2\}$
34. $\{17\}$
35. $\varnothing$
36. $\{36\}$

Objective 2

37. $\{4\}$
38. $\{1\}$
39. $\{2\}$
40. $\{3, 4\}$
41. $\{10\}$
42. $\varnothing$
43. $\{6\}$
44. $\varnothing$
45. $\{6\}$
46. $\{6\}$

47. $\left\{\dfrac{3}{4}\right\}$
48. $\{3, 4\}$
49. $\{3\}$
50. $\{-1\}$
51. $\{3\}$
52. $\{0, 8\}$
53. $\{3\}$
54. $\left\{\dfrac{3}{2}\right\}$
55. $\{7\}$

56. $\{3\}$
57. $\{-1\}$
58. $\{3\}$
59. $\{-2\}$
60. $\{7\}$
61. $\{1\}$
62. $\{-4, 0\}$
63. $\{5\}$
64. $\{11\}$
65. $\{0, -1\}$

66. $\{-7, -2\}$
67. $\{6\}$
68. $\{-11, 13\}$
69. $\{3\}$
70. $\{9\}$
71. $\{2\}$
72. $\{-9\}$

Objective 3

73. $\{3\}$
74. $\{28\}$
75. $\{-3\}$

76. $\varnothing$
77. $\{0\}$
78. $\{2\}$

79. $\{5\}$
80. $\{1\}$
81. $\{17\}$

82. $\{-9\}$
83. $\{33\}$
84. $\{3\}$

85. $\{-4\}$

86. $\{3\}$

87. $\{-31\}$

88. $\{21\}$

89. $\{0\}$

90. $\{0\}$

9.6   Mixed Exercises

91. $\varnothing$

92. $\{64\}$

93. $\left\{\dfrac{3}{2}, \dfrac{5}{2}\right\}$

94. $\{7\}$

95. $\{0\}$

96. $\{8\}$

97. $\{-5\}$

98. $\{-3\}$

99. $\left\{1, \dfrac{3}{2}\right\}$

100. $\{4\}$

9.7 Complex Numbers

Objective 1

1. $7i$

2. $6i$

3. $-10i$

4. $i\sqrt{6}$

5. $i\sqrt{22}$

6. $5i\sqrt{2}$

7. $-3i\sqrt{7}$

8. $2i\sqrt{30}$

9. $3i\sqrt{2}$

10. $-3i\sqrt{3}$

11. $2i\sqrt{15}$

12. $15i\sqrt{2}$

13. $-5i\sqrt{5}$

14. $-6i\sqrt{2}$

15. $3i\sqrt{11}$

16. $6i\sqrt{30}$

17. $-25i$

18. $-9i\sqrt{2}$

19. -6

20. $-3\sqrt{5}$

21. $-7\sqrt{3}$

22. $i\sqrt{42}$

23. $i\sqrt{21}$

24. $i\sqrt{30}$

25. -6

26. $-i\sqrt{105}$

27. 5

28. $2i\sqrt{2}$

29. $2i$

30. 3

31. 3

32. $4i$

33. 5

34. $6i$

35. $-\sqrt{21}$

36. $\dfrac{i\sqrt{70}}{5}$

Objectives 2

37. imaginary

38. imaginary

39. real

40. imaginary

41. real

42. imaginary

43. imaginary

44. real

45. imaginary

46. imaginary

Objective 3

47. $3 + 11i$

48. $8 + 6i$

49. $14 - 2i$

50. $7 + 6i$

51. 3

52. $-7 + 2i$

53. $-4 + i$

54. $5 + 3i$

55. $9 - 9i$

56. $-7 + 2i$

57. $-5 - i$

58. 1

59. $2 - 3i$

60. 10

61. $3\sqrt{3} - 4i\sqrt{2}$

62. $7 + 5i$

Objective 4

63. $11 + 13i$

64. $25 + 8i$

65. $44 + 12i$

66. 34

67. 53

68. $-13i$

69. $-8 + 6i$

70. $-1 - 2i\sqrt{6}$

71. $34 - 34i$

72. $7 + i$

Objective 5

73. $\dfrac{1}{5} + \dfrac{3}{5}i$

74. $\dfrac{37}{97} + \dfrac{38}{97}i$

75. $\dfrac{19}{17} + \dfrac{43}{17}i$

76. $-1 - 3i$

77. $\dfrac{10}{13} + \dfrac{11}{13}i$

78. $\dfrac{4}{5} - \dfrac{7}{5}i$

79. $\dfrac{15}{13} + \dfrac{16}{13}i$

80. $-\dfrac{3}{5} + \dfrac{6}{5}i$

81. $-\dfrac{1}{4} - \dfrac{1}{2}i$

82. $\dfrac{1}{5} + \dfrac{1}{5}i$

Objective 6

83. $-i$

84. i

85. 1

86. -1

87. $-i$

88. i

89. 1

90. -1

91. $-i$

92. 1

93. 1

94. $-i$

95. i

96. $-i$

97. $-i$

9.7   Mixed Exercises

98. $5i\sqrt{5}$

99. $2i$

100. i

101. i

102. $-7\sqrt{5}$

103. -8

104. $11 - 4i$

105. $13 + i$

106. $-14 + 10i$

107. $2 + 4i$

108. $17 + 7i$

109. 29

110. $\dfrac{9}{13} + \dfrac{20}{13}i$

111. $\dfrac{47}{25} - \dfrac{21}{25}i$

Chapter 10

QUADRATIC EQUATIONS, INEQUALITIES, AND FUNCTIONS

10.1 The Square Root Property and Completing the Square

Objective 1

1. $\{-9, 9\}$

2. $\{-3, 3\}$

3. $\left\{-\sqrt{7}, \sqrt{7}\right\}$

4. $\{-2, 2\}$

5. $\left\{-\sqrt{13}, \sqrt{13}\right\}$

6. $\left\{-\sqrt{2}, \sqrt{2}\right\}$

7. $\{-2, 2\}$

8. $\{-3, 3\}$

9. $\{-1, 1\}$

10. $\left\{-\sqrt{10}, \sqrt{10}\right\}$

Objective 2

11. $\{-9, 7\}$

12. $\left\{-\sqrt{3}-4, \sqrt{3}-4\right\}$

13. $\{-1, 11\}$

14. $\left\{\dfrac{-3+\sqrt{5}}{2}, \dfrac{-3-\sqrt{5}}{2}\right\}$

15. $\left\{-4, -\dfrac{2}{3}\right\}$

16. $\left\{\dfrac{1+\sqrt{6}}{7}, \dfrac{1-\sqrt{6}}{7}\right\}$

17. $\{0, -12\}$

18. $\left\{-\dfrac{7}{4}, -\dfrac{5}{4}\right\}$

19. $\{-3, 5\}$

20. $\{-3, 9\}$

21. $\left\{-1, \dfrac{7}{3}\right\}$

22. $\left\{-\dfrac{3}{2}, \dfrac{5}{2}\right\}$

Objective 3

23. $\{-3, -1\}$

24. $\left\{0, \dfrac{5}{2}\right\}$

25. $\left\{\dfrac{2+\sqrt{6}}{2}, \dfrac{2-\sqrt{6}}{2}\right\}$

26. $\{-4, -3\}$

27. $\left\{\dfrac{-1+3\sqrt{5}}{2}, \dfrac{-1-3\sqrt{5}}{2}\right\}$

28. $\left\{\dfrac{-2+\sqrt{7}}{3}, \dfrac{-2-\sqrt{7}}{3}\right\}$

29. $\left\{1-\sqrt{3}, 1+\sqrt{3}\right\}$

30. $\left\{\dfrac{1+\sqrt{29}}{4}, \dfrac{1-\sqrt{29}}{4}\right\}$

31. $\{-2, 0\}$

32. $\{3, 9\}$

33. $\{-1, 2\}$

34. $\{-2, -1\}$

35. $\{1, 8\}$

36. $\left\{2-\sqrt{13}, 2+\sqrt{13}\right\}$

Objective 4

37. $\{-i, i\}$

38. $\{-3i, 3i\}$

39. $\{-4i, 4i\}$

40. $\left\{ \dfrac{7-3i}{6}, \dfrac{7+3i}{6} \right\}$

43. $\left\{ -\dfrac{5}{2}i, \dfrac{5}{2}i \right\}$

46. $\left\{ \dfrac{-16-7i}{8}, \dfrac{-16+7i}{8} \right\}$

41. $\{-5i, 5i\}$

44. $\{-11i, 11i\}$

42. $\{-1-6i, -1+6i\}$

45. $\{-9i, 9i\}$

10.1   Mixed Exercises

47. $\left\{ -5+\sqrt{7}, -5-\sqrt{7} \right\}$

54. $\left\{ -\sqrt{10}, \sqrt{10} \right\}$

48. $\{-15, 15\}$

55. $\{-3, 0\}$

49. $\left\{ -3\sqrt{3}, 3\sqrt{3} \right\}$

56. $\left\{ \dfrac{-7+3\sqrt{5}}{2}, \dfrac{-7-3\sqrt{5}}{2} \right\}$

50. $\left\{ \dfrac{-4+\sqrt{7}}{2}, \dfrac{-4-\sqrt{7}}{2} \right\}$

57. $\left\{ \dfrac{-7+3\sqrt{17}}{4}, \dfrac{-7-3\sqrt{17}}{4} \right\}$

51. $\{3+i, 3-i\}$

52. $\left\{ \dfrac{-5+2i\sqrt{3}}{4}, \dfrac{-5-2i\sqrt{3}}{4} \right\}$

58. $\left\{ 1+i\sqrt{2}, 1-i\sqrt{2} \right\}$

59. $\{1, 4\}$

53. $\left\{ -\dfrac{10}{3}, \dfrac{10}{3} \right\}$

60. $\left\{ -\dfrac{3}{2}, 2 \right\}$

10.2 The Quadratic Formula

Objective 1

1. $\{1, 6\}$

2. $\{3, 9\}$

3. $\{-7, 2\}$

4. $\{-5, 8\}$

5. $\{4\}$

6. $\left\{\dfrac{3}{5}, 2\right\}$

7. $\left\{\dfrac{4}{3}, \dfrac{3}{2}\right\}$

8. $\{-4, 2\}$

9. $\left\{-\dfrac{3}{4}, \dfrac{3}{4}\right\}$

10. $\{2, 4\}$

11. $\left\{\dfrac{-5+3\sqrt{5}}{10}, \dfrac{-5-3\sqrt{5}}{10}\right\}$

12. $\left\{\dfrac{-1+\sqrt{2}}{2}, \dfrac{-1-\sqrt{2}}{2}\right\}$

13. $\left\{\dfrac{3+\sqrt{6}}{3}, \dfrac{3-\sqrt{6}}{3}\right\}$

14. $\left\{\dfrac{2+i\sqrt{2}}{2}, \dfrac{2-i\sqrt{2}}{2}\right\}$

15. $\left\{-\dfrac{1}{4}, 3\right\}$

16. $\left\{\dfrac{2}{3}, \dfrac{3}{2}\right\}$

Objective 2

17. D
18. D
19. D
20. A
21. D
22. C
23. B
24. D
25. B
26. C
27. C
28. A
29. C
30. A

10.2   Mixed Exercises

31. $\left\{1+i\sqrt{2}, 1-i\sqrt{2}\right\}$

32. $\left\{\dfrac{1+\sqrt{13}}{2}, \dfrac{1-\sqrt{13}}{2}\right\}$

33. $\{-2, 14\}$
34. $\{-2, 1\}$
35. A
36. C
37. A
38. A
39. B
40. D

10.3 Equations Quadratic in Form

Objective 1

1. $\{-3, 2\}$

2. $\left\{-\dfrac{5}{3}, 3\right\}$

3. $\left\{-\dfrac{7}{5}, 2\right\}$

4. $\left\{-\dfrac{7}{2}, -\dfrac{1}{3}\right\}$

5. $\left\{-3, -\dfrac{3}{2}\right\}$

6. $\left\{-7, \dfrac{5}{4}\right\}$

7. $\left\{-\dfrac{7}{2}, 4\right\}$

8. $\left\{\dfrac{2}{3}\right\}$

9. $\{-3, 5\}$

10. $\left\{-3, -\dfrac{1}{5}\right\}$

11. $\left\{\dfrac{9}{7}, 4\right\}$

12. $\left\{-\dfrac{35}{4}, -3\right\}$

13. $\left\{-7, \dfrac{5}{2}\right\}$

14. $\left\{-\dfrac{5}{2}, 1\right\}$

15. $\left\{\dfrac{8}{3}, 6\right\}$

16. $\{-1, 2\}$

Objective 2

17. 4 in. by 8 in.

18. 5 mph

19. 2 mph

20. 10 hr, 15 hr

21. 50 mph

22. 15 hr, 30 hr

23. 550 mph

24. 3 or $-\dfrac{40}{11}$

25. bike: 12 mph; hike: 2 mph

26. 5 or $\dfrac{1}{5}$

Objective 3

27. $\{2\}$

28. $\{2, 5\}$

29. $\left\{\dfrac{3}{2}\right\}$

30. $\{3, 5\}$

31. $\left\{\dfrac{3}{2}\right\}$

32. $\left\{\dfrac{1}{4}\right\}$

33. $\left\{\dfrac{7}{3}, 7\right\}$

34. $\left\{\dfrac{1}{4}, \dfrac{1}{3}\right\}$

35. $\left\{\dfrac{1}{16}, \dfrac{1}{9}\right\}$

36. $\{4\}$

Objective 4

37. $\{-4, -3, 3, 4\}$

38. $\left\{-1, -\dfrac{3}{4}, \dfrac{3}{4}, 1\right\}$

39. $\{-6, 14\}$

40. $\{-3, -2, 2, 3\}$

41. $\{-9, -7\}$

42. $\{0, 9\}$

43. $\{9\}$

44. $\left\{-\dfrac{1}{3}, \dfrac{1}{4}\right\}$

45. $\{-2, 2\}$

46. $\left\{-3, -\sqrt{7}, \sqrt{7}, 3\right\}$

47. $\left\{-\sqrt{5}, -\dfrac{1}{2}, \dfrac{1}{2}, \sqrt{5}\right\}$

48. $\left\{-\dfrac{17}{3}, -4\right\}$

49. $\left\{-3\sqrt{2}, -\sqrt{2}, \sqrt{2}, 3\sqrt{2}\right\}$

50. $\left\{-2\sqrt{5}, 2\sqrt{5}\right\}$

51. $\left\{-2\sqrt{2}, 2\sqrt{2}, -\sqrt{3}, \sqrt{3}\right\}$

52. $\left\{-\sqrt{10}, -\sqrt{2}, \sqrt{2}, \sqrt{10}\right\}$

53. $\left\{-\sqrt{6}, -2, 2, \sqrt{6}\right\}$

54. $\left\{-3, -\sqrt{3}, \sqrt{3}, 3\right\}$

55. $\{-1, 1\}$

56. $\left\{-\sqrt{3}, 0, \sqrt{3}\right\}$

57. $\left\{-\sqrt{3}, 0, \sqrt{3}\right\}$

58. $\left\{-10, -\dfrac{1}{6}, \dfrac{1}{6}, 10\right\}$

10.3 Mixed Exercises

59. 18.6 hr

60. 25.2 hr, 37.2 hr

61. 5.9 mph

62. $\left\{-\dfrac{7}{2}, \dfrac{1}{5}\right\}$

63. $\{3, 4\}$

64. $\left\{-7, -\dfrac{7}{2}\right\}$

65. $\{-3, 2\}$

66. $\left\{-\dfrac{4}{3}, 2\right\}$

67. $\{-2, -1, 4, 5\}$

68. $\{9\}$

69. $\left\{-2, -\dfrac{\sqrt{3}}{2}, \dfrac{\sqrt{3}}{2}, 2\right\}$

10.4 Formulas and Further Applications

Objective 1

1. $k = \dfrac{D^2}{h}$

2. $d = \dfrac{k^2 l^2}{F^2}$

3. $k = \dfrac{p^2 g}{l}$

4. $z = \dfrac{p\sqrt{6}}{y}$

5. $p = \dfrac{900a}{s^2}$

6. $c = \dfrac{(a-1)^2}{b}$

7. $t = \dfrac{\pm\sqrt{2gy}}{g}$

8. $t = \dfrac{\pm\sqrt{mxF}}{F}$

9. $x = \dfrac{\pm\sqrt{2Fk}}{k}$

Objective 2

10. 12 ft

11. 10 ft

12. south: 72 mi; east: 54 mi

13. north: 57 mi; west: 76 mi

14. 36 ft

15. 60 ft

16. 34 cm

17. 8 ft

18. 12 m

19. 25 ft

20. 48 cm

21. 24 in.

Objective 3

22. 3 m by 5 m

23. 12 in. by 16 in.

24. 3 cm

25. 7 in. by 11 in.

26. 12 in. by 15 in.

27. 8 in. by 5 in. by 4 in.

28. 9 in. by 14 in.

29. 16 cm by 10 cm

30. 2 ft

31. 3 ft

32. 2.5 in.

33. 2.5 ft

Objective 4

34. 17.1 hr

35. 1.4 min

36. 9.1 sec

37. 1.2 sec

10.4   Mixed Exercises

38. $t = \dfrac{p \pm \sqrt{p^2 + 4pq}}{2}$

39. $m = \dfrac{-x \pm \sqrt{x^2 + 4xy}}{2}$

40. $q = \dfrac{-k \pm k\sqrt{5}}{2p}$

41. $a = \dfrac{-c \pm c\sqrt{2}}{b}$

42. 4 ft

43. 15 cm

44. 13 in.

45. 1.5 ft

46. 27 items

10.5 Graphs of Quadratic Functions

Objectives 1 and 2

1. $(0, -2)$

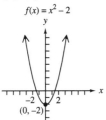

5. $(0, 2)$

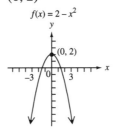

9. $(-3, -1)$

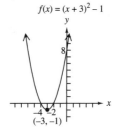

2. $(0, 2)$

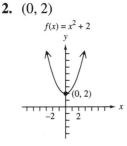

6. $(0, 5)$

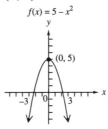

10. $(2, 1)$

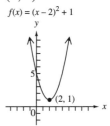

3. $(0, 3)$

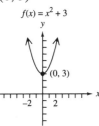

7. $(-2, 0)$

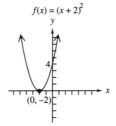

4. $(0, -4)$

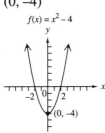

8. $(3, 0)$

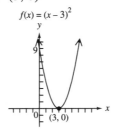

Objective 3

11. up; wider

12. up; narrower

13. down; narrower

14. down; narrower

15. up; wider

16. up; wider

17. down; narrower

18. down; wider

19. up; narrower

20. down; same

Objective 4

21. quadratic; negative

22. linear; positive

23. quadratic; positive

10.5   Mixed Exercises

24. (1, 0)

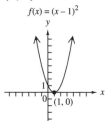

$f(x) = (x - 1)^2$

25. (−2, 3)

$f(x) = (x + 2)^2 + 3$

26. (3, −1)

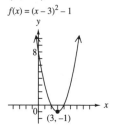

$f(x) = (x - 3)^2 - 1$

27. (−3, 0)

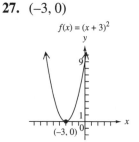

$f(x) = (x + 3)^2$

28. up; narrower; (0, 0)

29. down, same; (0, 2)

30. up; wider; (1, 0)

31. up; narrower; (−1, −2)

32. down; wider; (−3, 0)

33. up; narrower; (3, 1)

10.6 More About Parabolas; Applications

Objectives 1 and 2

1. $(-3, 1)$

$f(x) = x^2 + 6x + 10$

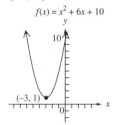

2. $(3, -5)$

$f(x) = x^2 - 6x + 4$

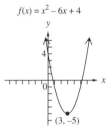

3. $(4, 6)$

$f(x) = -x^2 + 8x - 10$

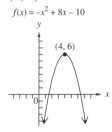

4. $\left(\dfrac{3}{2}, -\dfrac{1}{4}\right)$

$f(x) = x^2 - 3x + 2$

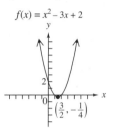

5. $(-1, -1)$

$f(x) = 3x^2 + 6x + 2$

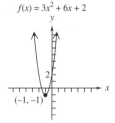

6. $(1, 3)$

$f(x) = -2x^2 + 4x + 1$

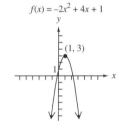

7. $(-2, 1)$

$f(x) = \frac{1}{2}x^2 + 2x + 3$

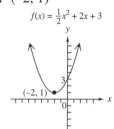

8. $(-2, -2)$

$f(x) = \frac{5}{4}x^2 + 5x + 3$

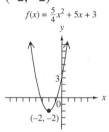

Objective 3

9. 0

10. 0

11. 0

12. 2

13. 0

14. 2

15. 2

16. 0

17. 1

18. 2

Objective 4

19. 25 units; $3650

20. 25 units; $1000

21. 50 pots; $200

22. 64 ft; 2 sec

23. 16 ft; 1 sec

24. 286 ft; $\dfrac{3}{2}$ sec

25. 256 ft; $\dfrac{5}{2}$ sec

26. 24 (a square)

27. 32 (a square)

28. 6 (a square)

Objective 5

29.   domain: $[0, \infty)$
range: $(-\infty, \infty)$

$x = 2y^2$

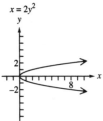

30.   domain: $(-\infty, 0]$
range: $(-\infty, \infty)$

$x = -2y^2$

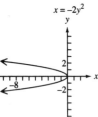

31.   domain: $(-\infty, 2]$
range: $(-\infty, \infty)$

$x = -y^2 + 2$

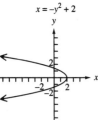

32.   domain: $[-3, \infty)$
range: $(-\infty, \infty)$

$x = y^2 - 3$

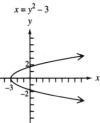

33.   domain: $[0, \infty)$
range: $(-\infty, \infty)$

$x = y^2 + 4y + 4$

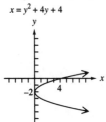

34.   domain: $(-\infty, 0]$
range: $(-\infty, \infty)$

$x = -y^2 + 4y - 4$

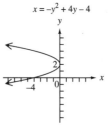

35.   domain: $[3, \infty)$
range: $(-\infty, \infty)$

$x = y^2 - 4y + 7$

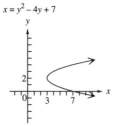

36.   domain: $[-1, \infty)$
range: $(-\infty, \infty)$

$3x = y^2 - 6y + 6$

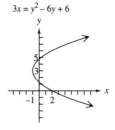

10.6 Mixed Exercises

37. $(-3, -1)$

$f(x) = -\frac{1}{3}x^2 - 2x - 4$

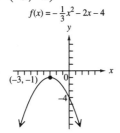

38. $(-4, -3)$

$x = y^2 + 6y + 5$

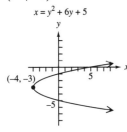

39. $(-1, -3)$

$x = -y^2 - 6y - 10$

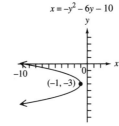

40. $\left(-1, -\dfrac{5}{2}\right)$

$f(x) = 2x^2 + 4x - \frac{1}{2}$

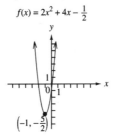

41. 2

42. 0

43. 8 units; $98

44. 4 (square)

45. 46 and 46

10.7 Quadratic and Rational Inequalities

Objective 1

1. $(-\infty, -3] \cup [2, \infty)$

2. $(-3, 2)$

3. $(-2, 5)$

4. $(-\infty, -3] \cup [2, \infty)$

5. $(-\infty, -4) \cup (-3, \infty)$

6. $[-1, 2]$

7. $\left(-1, \dfrac{3}{2}\right)$

8. $\left(-\infty, -\dfrac{3}{2}\right) \cup (4, \infty)$

9. $\left(-\infty, -\dfrac{2}{3}\right) \cup \left(-\dfrac{1}{2}, \infty\right)$

10. $\left(-\infty, -\dfrac{3}{2}\right) \cup \left(\dfrac{1}{4}, \infty\right)$

11. $(-\infty, \infty)$

12. $\varnothing$

13. $\varnothing$

14. $(-\infty, \infty)$

15. $\varnothing$

16. $\varnothing$

Objective 2

17. $(-\infty, -4] \cup [-1, 2]$

18. $(-\infty, -2) \cup (1, 2)$

19. $(-\infty, -5] \cup [-3, 1]$

20. $[-2, 1] \cup [3, \infty)$

21. $(-\infty, -3] \cup [-1, 4]$

22. $(2, 4) \cup (6, \infty)$

23. $\left(-\infty, -\dfrac{3}{2}\right] \cup \left[-\dfrac{1}{3}, \dfrac{1}{2}\right]$

24. $\left(-\dfrac{1}{4}, \dfrac{1}{6}\right) \cup \left(\dfrac{7}{3}, \infty\right)$

26. $(-\infty, -1) \cup \left(1, \dfrac{7}{3}\right)$

25. $\left[\dfrac{3}{4}, \dfrac{10}{3}\right] \cup \left[\dfrac{7}{2}, \infty\right)$

Objective 3

27. $(-\infty, 1) \cup [8, \infty)$

36. $\left(-\infty, -\dfrac{19}{9}\right] \cup \left(-\dfrac{5}{3}, \infty\right)$

28. $(-\infty, -3) \cup [-1, \infty)$

37. $\left(-\infty, \dfrac{1}{4}\right] \cup \left(\dfrac{3}{2}, \infty\right)$

29. $(-1, 4]$

38. $\left[0, \dfrac{3}{4}\right)$

30. $\left(-2, -\dfrac{2}{3}\right)$

31. $\left(-\dfrac{5}{3}, -1\right)$

39. $\left[-\dfrac{3}{2}, -1\right)$

32. $\left(-\infty, \dfrac{5}{2}\right] \cup (3, \infty)$

40. $\left(2, \dfrac{8}{3}\right]$

33. $(-\infty, -2)$

34. $(-4, \infty)$

41. $(-\infty, -2] \cup \left(-\dfrac{1}{3}, \infty\right)$

35. $\left(\dfrac{2}{3}, \dfrac{8}{3}\right]$

42. $(-\infty, 3) \cup [8, \infty)$

43. $[-2, 3)$

44. $\left(\dfrac{15}{13}, \dfrac{5}{3}\right)$

10.7 Mixed Exercises

45. $\left[\dfrac{1}{3}, \dfrac{2}{5}\right]$

46. $\left(-\infty, -\dfrac{2}{3}\right] \cup \left[\dfrac{5}{2}, \infty\right)$

47. $\left[-\dfrac{3}{2}, \dfrac{3}{2}\right]$

48. $\left(-\infty, -\dfrac{3}{4}\right] \cup \left[\dfrac{3}{4}, \infty\right)$

49. $\left(-\infty, -\dfrac{1}{2}\right) \cup \left(\dfrac{1}{4}, \infty\right)$

50. $\left(-3, \dfrac{1}{6}\right)$

51. $[-3, -1] \cup [2, \infty)$

52. $(-\infty, -2) \cup (2, 3)$

53. $(-\infty, -1) \cup \left(\dfrac{1}{2}, 3\right)$

54. $\left[\dfrac{3}{4}, 1\right] \cup [3, \infty)$

55. $\left(-\dfrac{3}{2}, \dfrac{1}{3}\right) \cup (2, \infty)$

56. $\left(-\infty, -\dfrac{5}{2}\right] \cup \left[-1, \dfrac{4}{5}\right]$

57. $\left[-\dfrac{3}{2}, 0\right)$

58. $(-\infty, 0) \cup \left[\dfrac{1}{2}, \infty\right)$

59. $\left(0, \dfrac{3}{5}\right]$

60. $(-\infty, 0) \cup \left[\dfrac{3}{8}, \infty\right)$

61. $\varnothing$

62. $(-\infty, \infty)$

Chapter 11

EXPONENTIAL AND LOGARITHMIC FUNCTIONS

11.1 Inverse Functions

Objective 1

1. not one-to-one

2. not one-to-one

3. $\{(-1, 2), (1, -2), (3, 1), (-3, -1)\}$

4. $\{(-3, 6), (-2, 4), (-1, 2), (0, 0)\}$

5. not one-to-one

6. not one-to-one

7. $\{(1, -1), (2, -2), (3, -3)\}$

8. $\{(1, -3), (2, -2), (3, -1), (4, 0)\}$

9. $\{(2, 3), (-2, -3), (3, 2), (-3, -2)\}$

10. not one-to-one

Objective 2

11. one-to-one

12. not one-to-one

13. not one-to-one

14. not one-to-one

15. one-to-one

16. one-to-one

17. not one-to-one

18. one-to-one

Objective 3

19. $f^{-1}(x) = \dfrac{x+5}{2}$

20. $f^{-1}(x) = \dfrac{x+5}{3}$

21. not one-to-one

22. not one-to-one

23. $f^{-1}(x) = x^2 + 1; \ x \geq 0$

24. $f^{-1}(x) = \dfrac{x^2}{12}; \ x \geq 0$

25. $f^{-1}(x) = \sqrt[3]{x+1}$

26. $f^{-1}(x) = \dfrac{\sqrt[3]{4x+12}}{2}$

27. not one-to-one

28. $f^{-1}(x) = \dfrac{3+x}{x}$

Objective 4

29.

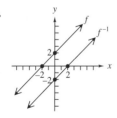

30.

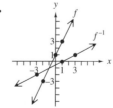

31. not one-to-one

34. not one-to-one

32.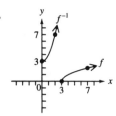

35.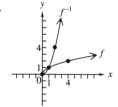

33. not one-to-one

36. not one-to-one

11.1 Mixed Exercises

37. $\{(5, 3), (9, 2), (7, 4)\}$

45.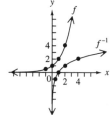

38. $\{(0, 0), (1, 1), (-1, -1), (2, 2), (-2, -2)\}$

39. not one-to-one

40. $\{(-1, -3), (0, -2), (1, -1), (2, 0)\}$

41. $f^{-1}(x) = \dfrac{4-x}{2}$

46.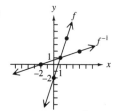

42. $f^{-1}(x) = x^2 - 2; \; x \geq 0$

43. not one-to-one

44. $f^{-1}(x) = \sqrt[3]{x+5}$

11.2 **Exponential Functions**

Objective 1

1. exponential function

2. not an exponential function

3. not an exponential function

4. exponential function

5. not an exponential function

6. not an exponential function

7. exponential function

8. exponential function

9. not an exponential function

10. not an exponential function

Objective 2

11.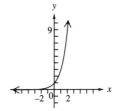

13.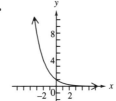

15.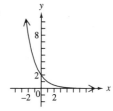

12.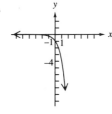

14.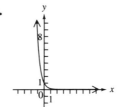

16.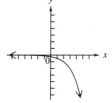

Objective 3

17. $\left\{\dfrac{3}{2}\right\}$

18. $\left\{\dfrac{2}{3}\right\}$

19. $\{2\}$

20. $\left\{\dfrac{3}{4}\right\}$

21. $\left\{\dfrac{1}{2}\right\}$

22. $\{1\}$

23. $\left\{\dfrac{1}{2}\right\}$

24. $\left\{\dfrac{3}{4}\right\}$

25. $\left\{\dfrac{1}{2}\right\}$

26. $\left\{-\dfrac{1}{2}\right\}$

Objective 4

27. 515 geese

28. 7.5 in.

29. 3650 bacteria

30. 35,000

31. 1 g

32. 200 bacteria

33. 1000 bacteria

34. 256,000

11.2   **Mixed Exercises**

35. $\left\{\dfrac{3}{2}\right\}$

36. $\left\{-\dfrac{5}{4}\right\}$

37. $\left\{\dfrac{1}{4}\right\}$

38. $\{1\}$

39. $\{-3\}$

40. $\{-2\}$

41.

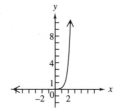

42.

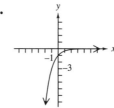

43. $27,048.14

44. 250,000 barrels

11.3 Logarithmic Functions

Objective 1

1. 3

2. 2

3. −1

4. $\dfrac{1}{2}$

5. −2

6. −4

7. $\dfrac{1}{2}$

8. $\dfrac{1}{2}$

9. −2

10. $\dfrac{3}{4}$

Objective 2

11. $\log_3 9 = 2$

12. $\log_5 \sqrt[3]{5} = \dfrac{1}{3}$

13. $4^{-2} = \dfrac{1}{16}$

14. $16^{1/4} = 2$

15. $\log_{10} \dfrac{1}{100} = -2$

16. $5^2 = 25$

17. $9^{1/2} = 3$

18. $\log_9 3 = \dfrac{1}{2}$

19. $10^{-3} = .001$

20. $\log_2 \dfrac{1}{128} = -7$

Objective 3

21. $\{6\}$

22. $\left\{\dfrac{1}{5}\right\}$

23. $\{5\}$

24. $\{3\}$

25. $\left\{\dfrac{1}{32}\right\}$

26. $\{2\}$

27. $\{10\}$

28. $\{4\}$

29. $\{1\}$

30. $\{1\}$

Objective 4

31.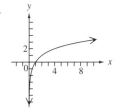

33.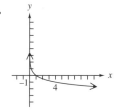

35.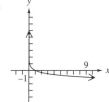

37.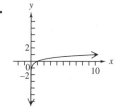

32.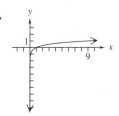

34.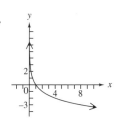

36.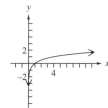

38.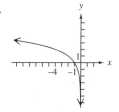

Objective 5

39. 20 squirrels

40. $600

41. 1200 cm/sec

42. 110 decibels

43. 8 students

44. 40 fish

45. 12 squirrels

46. 200 applicants

47. 335,000 items

48. 100 mites

11.3   **Mixed Exercises**

49. $\log_{1/2} 8 = -3$

50. $5^{-4} = .0016$

51. $\left\{ \dfrac{3}{5} \right\}$

52. $\{81\}$

53. $\{0\}$

54. $\{16\}$

55.

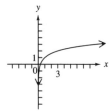

56. 239 foxes

57. 16 fish

58. $100,000

11.4 Properties of Logarithms

Objective 1

1. $\log_3 6 + \log_3 5$

2. $\log_2 5 + \log_2 3$

3. $\log_7 5 + \log_7 m$

4. $\log_2 6 + \log_2 x + \log_2 y$

5. $1 + \log_6 r$

6. $\log_3 2 + \log_3 p$

7. $\log_4 21$

8. $\log 12$

9. $\log_7 66 y^3$

10. $\log_7 120 r^5$

Objective 2

11. $\log_2 7 - \log_2 9$

12. $\log_4 5 - \log_4 8$

13. $\log_3 m - \log_3 n$

14. $\log p - \log r$

15. $\log_6 k - \log_6 3$

16. $\log_3 10 - \log_3 x$

17. $3 - \log_2 m$

18. $1 - \log_5 x$

19. $\log_2 \dfrac{7q^2}{5}$

20. $\log 3x$

21. $\log_7 \dfrac{3}{5r^4}$

22. $\log_9 \dfrac{2}{y^2}$

Objective 3

23. $2 \log_5 3$

24. $3 \log_3 4$

25. $3 \log_2 5$

26. $7 \log_m 2$

27. $\dfrac{1}{2} \log_b 5$

28. $\dfrac{1}{3} \log_3 7$

29. $\dfrac{1}{2}$

30. $\dfrac{1}{3}$

31. 1

32. 1

Objective 4

33. $2 + 3 \log_2 p$

34. $2 + 3 \log_3 x$

35. $\dfrac{1}{3}(\log_a 2 + \log_a k)$

36. $\log_b 2 + \log_b r - \log_b (r-1)$

37. $2 - \log_2 3$

38. $\log_3 5 - 2$

39. $\log_5 7 + 3 \log_5 m - \log_5 8 - \log_5 y$

40. $\log_7 8 + 7 \log_7 r - \log_7 3 - 3 \log_7 a$

41. $\log 14 x^2$

42. $\log_a 8 r^3$

43. $\log_b \dfrac{3q}{2p}$

44. $\log \dfrac{4k}{3j}$

45. $\log_4 \dfrac{5}{y}$

46. $\log_6 \dfrac{7}{m}$

47. 0

48. 0

11.4   Mixed Exercises

49. $3 + \log_2 p$

50. $\dfrac{3}{2}$

51. $1 - \log_4 9$

52. $4 \log_5 k$

53. 2

54. $\dfrac{1}{2}\log_5 3 + \dfrac{1}{2}\log_5 p$

55. $\log_7 3 - 1$

56. $\log_4 3 + \log_4 m - \log_4(m+2)$

57. 1

58. $\log_3 10q^4$

59. $\log_5 16y^2$

60. 1

61. $\log_5 \dfrac{1}{m^4}$ or $-\log_5 m^4$

62. $\log_2 \dfrac{1}{y}$ or $-\log_2 y$

11.5 Common and Natural Logarithms

Objective 1

1. 1.7576

6. 2.9025

11. 1.3424

16. −3

2. .9031

7. −4.4970

12. −.4558

17. 6.0133

3. 2.9262

8. 1.7853

13. 2.8848

18. .6022

4. −1.0390

9. −3.0186

14. 3.7395

5. 5.4472

10. 4.9401

15. −3.0814

Objective 2

19. 5.6

22. 8.8

25. 1.3×10^{-3}

28. 5.0×10^{-2}

20. 7.3

23. 4.2

26. 4.0×10^{-4}

29. 3.2×10^{-7}

21. 6.7

24. 10.1

27. 6.3×10^{-6}

30. 6.3×10^{-11}

Objective 3

31. −2.1203

34. 6.0591

37. 1.3863

40. 4.3347

32. 4.6052

35. −4.3428

38. −2.2828

41. −4.6052

33. 1.7918

36. 4.2341

39. 6.7731

42. −6.1469

Objective 4

43. 600

46. 733

49. 40.7 g

52. 11,600

44. 618

47. 100 g

50. 16.5 g

53. 12,200

45. 663

48. 74.1 g

51. 10,000

54. 33,200

11.5   Mixed Exercises

55. −1.0286

57. −7.1234

59. 8.4

61. 3814 termites

56. 3.9120

58. 4.7795

60. 1.6×10^{-5}

62. 20,000

11.6 Exponential and Logarithmic Equations; Further Applications

Objective 1

1. {.488}   **3.** {1.232}   **5.** {4.292}   **7.** {3.936}   **9.** {−.183}

2. {.517}   **4.** {−1.395}   **6.** {3.472}   **8.** {−1.339}   **10.** {1.380}

Objective 2

11. {5}

12. {3}

13. $\left\{\dfrac{1}{4}\right\}$ **15.** $\left\{-\dfrac{5}{6}\right\}$ **17.** $\left\{\sqrt[4]{10}\right\}$ **19.** $\left\{\dfrac{9}{8}\right\}$

18. {625}

14. {3}   **16.** {2}   **20.** {1}

Objective 3

21. $1259.71   **23.** $12,905.41   **25.** $5540.08

22. $40,262.75   **24.** $6391.88   **26.** $111,295.13

Objective 4

27. 8.25 g   **28.** 29 yr   **29.** 264 g   **30.** 22 yr

Objective 5

31. .6131   **34.** 3.2266   **37.** 3.8074   **40.** −1.7297   **43.** .4150

32. 3.3219   **35.** 4.5110   **38.** 3.4130   **41.** −1.2770   **44.** 1.6309

33. 1.7124   **36.** 1.1887   **39.** −2.5850   **42.** −3.9694

11.6   Mixed Exercises

45. {1.7227}   **48.** {4.1885}   **51.** $114,237.51   **54.** 7 yr

46. {−.6309}   **49.** {1000}   **52.** $68,098.05   **55.** 1.1950

47. {3.3802}   **50.** {1}   **53.** 18 yr   **56.** 1.5

Chapter 12

NONLINEAR FUNCTIONS, CONIC SECTIONS, AND NONLINEAR SYSTEMS

12.1 Additional Graphs of Functions; Composition

Objective 1

1. $f(x) = |x - 2| + 3$

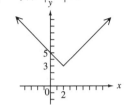

4. $f(x) = -|x + 3| - 2$

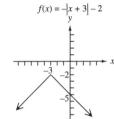

7. $f(x) = |x - 3| - 2$

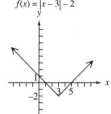

2. $f(x) = \sqrt{x + 3}$

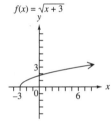

5. $f(x) = \sqrt{5 - x}$

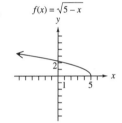

8. $f(x) = \sqrt{x} + 3$

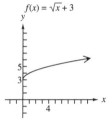

3. $f(x) = \frac{1}{x - 1}$

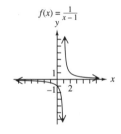

6. $f(x) = \frac{1}{x} + 3$

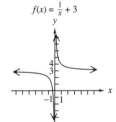

Objective 3

9. 12

10. 28

11. 38

12. 23

13. 4

14. $9x^2 + 12x + 7$

15. $3x + 14$

16. $x^2 + 8x + 19$

17. $3x^2 + 11$

18. $3x + 6$

12.1 Mixed Exercises

19. $f(x) = |x - 4| + 2$

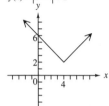

20. $f(x) = -\sqrt{x - 3} - 3$

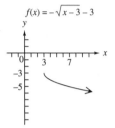

21. $f(x) = \frac{1}{x - 3}$

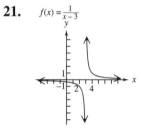

22. $f(x) = -|x - 3| + 3$

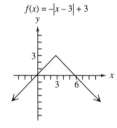

23. $f(x) = \sqrt{4 - x^2}$

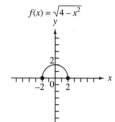

24. $f(x) = -\sqrt{1 - x^2}$

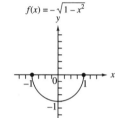

25. 7

26. 10

27. $x^2 + 3$

28. $x^2 - 6x + 15$

12.2 The Circle and the Ellipse

Objective 1

1. $(x+3)^2 + (y-2)^2 = 25$

2. $(x-1)^2 + (y-4)^2 = 4$

3. $x^2 + (y-5)^2 = 9$

4. $(x-6)^2 + (y-2)^2 = 9$

5. $(x+5)^2 + (y-4)^2 = 16$

6. $(x-7)^2 + (y-1)^2 = 4$

7. $(x-3)^2 + (y+4)^2 = 25$

8. $(x-2)^2 + (y-2)^2 = 36$

9. $(x-1)^2 + (y-3)^2 = 25$

10. $(x+2)^2 + (y+2)^2 = 9$

Objective 2

11. $(2, -4)$; 3

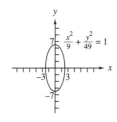

$x^2 + y^2 - 4x + 8y + 11 = 0$

12. $(-3, 2)$; 1

$x^2 + y^2 + 6x - 4y + 12 = 0$

13. $(3, -5)$; 8

14. $(2, 1)$; 6

15. $(-2, -3)$; 4

16. $(5, -6)$; 3

17. $(4, 1)$; $\sqrt{2}$

18. $(2, -4)$; 3

19. $(2, -1)$; $\sqrt{7}$

20. $(-5, -2)$; 6

Objectives 3 and 4

21.

$\dfrac{x^2}{9} + \dfrac{y^2}{49} = 1$

22.

$\dfrac{x^2}{25} + \dfrac{y^2}{4} = 1$

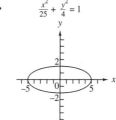

23. $\frac{x^2}{25} + \frac{y^2}{36} = 1$

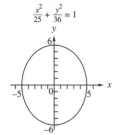

24. $\frac{x^2}{4} + \frac{y^2}{9} = 1$

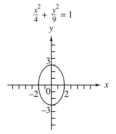

25. $\frac{x^2}{16} + \frac{y^2}{25} = 1$

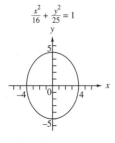

26. $\frac{x^2}{36} + \frac{y^2}{9} = 1$

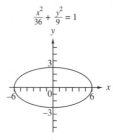

27. $\frac{x^2}{25} + \frac{y^2}{64} = 1$

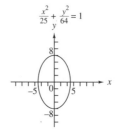

28. $\frac{x^2}{4} + \frac{y^2}{16} = 1$

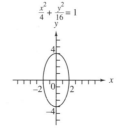

12.2   Mixed Exercises

29. $(x+2)^2 + (y+4)^2 = 25$

30. $x^2 + (y-3)^2 = 2$

31. $(-4, -2); 7$

32. $(3, -2); \dfrac{3}{2}$

33. $\frac{x^2}{16} + \frac{y^2}{49} = 1$

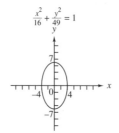

34. $\frac{x^2}{25} + \frac{y^2}{81} = 1$

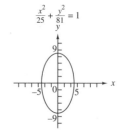

12.3   The Hyperbola and Other Functions Defined by Radicals

Objectives 1 and 2

1. $\frac{x^2}{9} - \frac{y^2}{16} = 1$

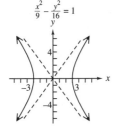

4. $\frac{y^2}{25} - \frac{x^2}{16} = 1$

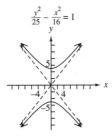

7. $\frac{x^2}{25} - \frac{y^2}{4} = 1$

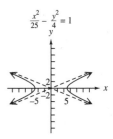

2. $\frac{x^2}{25} - \frac{y^2}{9} = 1$

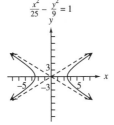

5. $\frac{x^2}{36} - \frac{y^2}{49} = 1$

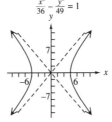

8. $\frac{x^2}{25} - \frac{y^2}{81} = 1$

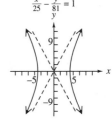

3. $\frac{y^2}{4} - \frac{x^2}{9} = 1$

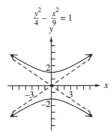

6. $\frac{y^2}{4} - \frac{x^2}{4} = 1$

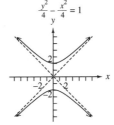

Objective 3

9. hyperbola

10. ellipse

11. parabola

12. hyperbola

13. hyperbola

14. parabola

15. circle

16. circle

17. circle

18. ellipse

Objective 4

19.

$f(x) = \sqrt{36 - x^2}$

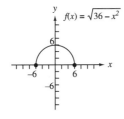

20.

$f(x) = \sqrt{25 - x^2}$

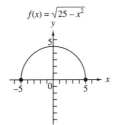

21.

$f(x) = -\sqrt{4 - x^2}$

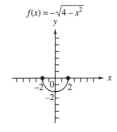

22. $f(x) = -\sqrt{9 - x^2}$

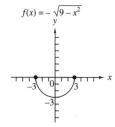

24. $f(x) = -3\sqrt{1 + x^2/25}$

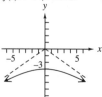

26. $f(x) = -5\sqrt{1 - x^2/9}$

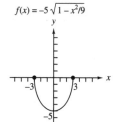

23. $f(x) = \sqrt{1 + x^2/4}$

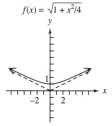

25. $f(x) = \sqrt{9 - 9x^2}$

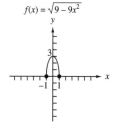

12.3   Mixed Exercises

27. parabola

29. circle

31. hyperbola

28. circle

30. parabola

32. parabola

12.4 Nonlinear Systems of Equations

Objective 1

1. $\left\{(4, -1), \left(\dfrac{16}{5}, -\dfrac{13}{5}\right)\right\}$

2. $\{(12, -17), (2, 3)\}$

3. $\left\{\left(\dfrac{1}{4}, \dfrac{3}{2}\right), (-1, 1)\right\}$

4. $\left\{(-2, 1), \left(-\dfrac{50}{17}, -\dfrac{31}{17}\right)\right\}$

5. $\{(5, 2), (-1, -4)\}$

6. $\{(0, 0), (-6, -2)\}$

7. $\{(3, -2), (-2, 3)\}$

8. $\{(3, 8), (-4, -6)\}$

9. $\left\{\left(\dfrac{5}{2}, -4\right), (2, -5)\right\}$

10. $\{(2, 5), (5, 2)\}$

Objective 2

11. $\{(1, 3), (1, -3), (-1, 3), (-1, -3)\}$

12. $\{(5, 2), (5, -2), (-5, 2), (-5, -2)\}$

13. $\{(2, 3), (2, -3), (-2, 3), (-2, -3)\}$

14. $\{(2, 1), (2, -1), (-2, 1), (-2, -1)\}$

15. $\{(3, 1), (3, -1), (-3, 1), (-3, -1)\}$

16. $\{(2, 3), (2, -3), (-2, 3), (-2, -3)\}$

17. $\{(1, 1), (1, -1), (-1, 1), (-1, -1)$

18. $\left\{\left(\dfrac{\sqrt{70}}{10}, \dfrac{3\sqrt{10}}{5}\right), \left(\dfrac{\sqrt{70}}{10}, \dfrac{-3\sqrt{10}}{5}\right), \left(\dfrac{-\sqrt{70}}{10}, \dfrac{3\sqrt{10}}{5}\right), \left(\dfrac{-\sqrt{70}}{10}, \dfrac{-3\sqrt{10}}{5}\right)\right\}$

19. $\{(2, 1), (2, -1), (-2, 1), (-2, -1)\}$

20. $\{(2, 0), (-2, 0)\}$

Objective 3

21. $\{(6, 1), (1, 6), (-6, -1), (-1, -6)\}$

22. $\{(2, 1), (-2, -1), (i, -2i), (-i, 2i)\}$

23. $\left\{(1, 1), (-1, -1), \left(\sqrt{3}, \dfrac{\sqrt{3}}{3}\right), \left(-\sqrt{3}, -\dfrac{\sqrt{3}}{3}\right)\right\}$

24. $\{(4, 3), (-4, -3), (3i, -4i), (-3i, 4i)\}$

25. $\{(2, -3), (-2, 3), (3, -2), (-3, 2)\}$

26. $\{(1, -4), (-1, 4), (4, -1), (-4, 1)\}$

27. $\left\{(3, -2), (-3, 2), \left(\dfrac{2\sqrt{6}}{3}, -\dfrac{3\sqrt{6}}{2}\right), \left(-\dfrac{2\sqrt{6}}{3}, \dfrac{3\sqrt{6}}{2}\right)\right\}$

28. $\left\{ (2, 1), (-2, -1), \left(\sqrt{2}, \sqrt{2}\right), \left(-\sqrt{2}, -\sqrt{2}\right) \right\}$

29. $\{(2, 2), (-2, -2), (2i, -2i), (-2i, 2i)\}$

30. $\left\{ (3, -1), (-3, 1), \left(\dfrac{\sqrt{2}}{2}, -3\sqrt{2}\right), \left(-\dfrac{\sqrt{2}}{2}, 3\sqrt{2}\right) \right\}$

12.4   Mixed Exercises

31. $\left\{ \left(\dfrac{3}{2}, \dfrac{1}{2}\right), \left(-\dfrac{3}{2}, -\dfrac{1}{2}\right) \right\}$

32. $\left\{ \left(2, \dfrac{1}{2}\right), (1, 1) \right\}$

33. $\left\{ \left(\sqrt{3}, 0\right), \left(-\sqrt{3}, 0\right) \right\}$

34. $\{(2, 1), (-2, -1), (i, -2i), (-i, 2i)\}$

35. $\{(3, 9), (-2, 4)\}$

36. $\{(2, 1), (2, -1), (-2, 1), (-2, -1)\}$

37. $\{(-2, 1), (2, -1), (i, 2i), (-i, -2i)\}$

38. $\left\{ \left(\dfrac{3+\sqrt{5}}{2}, \sqrt{5}\right), \left(\dfrac{3-\sqrt{5}}{2}, -\sqrt{5}\right) \right\}$

39. $\left\{ \left(\dfrac{i\sqrt{10}}{2}, -i\sqrt{10}\right), \left(\dfrac{-i\sqrt{10}}{2}, i\sqrt{10}\right), \left(\sqrt{5}, \sqrt{5}\right), \left(-\sqrt{5}, -\sqrt{5}\right) \right\}$

40. $\left\{ \left(\sqrt{2}, -\sqrt{2}\right), \left(-\sqrt{2}, \sqrt{2}\right) \right\}$

12.5 Second-Degree Inequalities and Systems of Inequalities

Objective 1

1. $x \geq y^2$

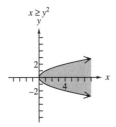

4. $25y^2 \leq 100 - 4x^2$

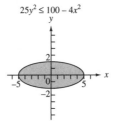

7. $x \leq 2y^2 + 8y + 9$

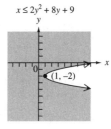

2. $y^2 \geq 9 - x^2$

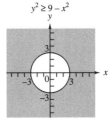

5. $x^2 + 4y^2 > 4$

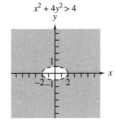

8. $4y^2 \geq 196 + 49x^2$

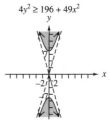

3. $16x^2 < 9y^2 + 144$

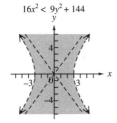

6. $y \geq x^2 - 4$

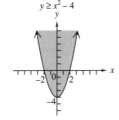

Objective 2

9. $-x + y > 2$
$3x + y > 6$

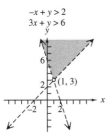

11. $x - 2y \geq -6$
$x + 4y \geq 12$

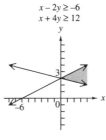

13. $9x^2 + 16y^2 < 144$
$y^2 - x^2 > 4$

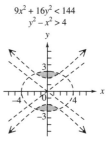

10. $x + y > -2$
$2x - y \leq -4$

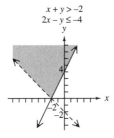

12. $x^2 + y^2 \leq 25$
$3x - 5y > -15$

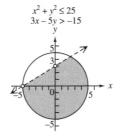

14. $x^2 + y^2 \leq 16$
$y \leq x^2 - 4$

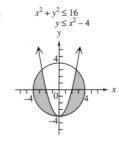

12.5   **Mixed Exercises**

15. $7x^2 \le 42 - 6y^2$

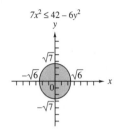

17. $x^2 > 9 - y^2$
$x \le 0$ and $y \ge 0$

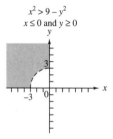

19. $4y + x^2 < 0$
$x \ge 0$

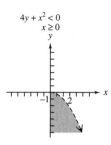

16. $9x^2 + 64y^2 \le 576$
$x \ge 0$

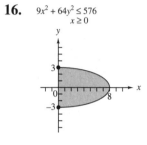

18. $x^2 - y^2 \le 16$
$y \ge 0$

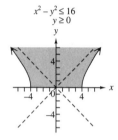

20. $x^2 + 4y^2 \le 36$
$-5 < x < 2$ and $y \ge 0$

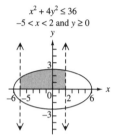